XXII Porch D

THE COSMIC HOLOGRAM

"The emerging scientific revolution in addressing the mind-body question and the deeper nature of reality is fuelled by some of the most profound revelations from the leading edge of scientific investigations, much of which is discussed in Jude Currivan's thoughtful and insightful book *The Cosmic Hologram*. Jude's formal training in physics combined with her transcendental spiritual experiences beginning at an early age set the stage for a fascinating tour of the fundamental underpinnings of reality."

EBEN ALEXANDER, M.D., NEUROSURGEON AND AUTHOR OF *PROOF OF HEAVEN*

"*The Cosmic Hologram* is a wonderfully important and very wise book. . . . the issues Jude Currivan, Ph.D., addresses will determine our fate as a viable species. . . . no one can consider herself an educated, responsible citizen in the twenty-first century who is not knowledgeable about these vital concerns. Written in elegant, clear, empathic prose, *The Cosmic Hologram* belongs at the top of your stack.

LARRY DOSSEY, M.D., PHYSICIAN, LECTURER, AND AUTHOR OF 11 BOOKS, INCLUDING *ONE MIND* AND THE NEW YORK TIMES BESTSELLER *HEALING WORDS*

"A stimulating and deeply inspiring view of the cosmos, consciousness, and the role of human perceptions in shaping what we call 'reality.' In the end, an offering of hope that during this Consciousness Revolution we can mold that reality into a compassionate and regenerative future."

JOHN PERKINS, COFOUNDER OF THE PACHAMAMA ALLIANCE AND AUTHOR OF *THE NEW CONFESSIONS OF AN ECONOMIC HIT MAN*

"*The Cosmic Hologram* is as elegant in its vision of a hologramatic universe as it is profound in its assertion that humanity is an integral part of it. We *are* the universe. Jude Currivan has thus pointed to the future by giving scientific description and credibility to the most ancient awareness of our species, that we are inside and not separate from all that is, that we are in fact shaping life even as it shapes us. This book is the new story for a planetary civilization."

Jim Garrison, Ph.D., president of Ubiquity University

"We highly recommend this timely, leading-edge book as it brings together consciousness with scientifically based evidence in a way we can all understand to explain the 'revolution of consciousness' now taking place!"

J. J. Hurtak, Ph.D., and Desiree Hurtak, Ph.D., social scientists, futurists, and founders of The Academy for Future Science

"Jude Currivan, Ph.D., lucidly develops the exciting and revolutionary insight and provides the transformational evidence that we live within a universe that is a high resolution, dynamically regenerated hologram. *The Cosmic Hologram* masterfully brings together extensive scientific knowledge, spiritual insight, and embodied wisdom in a seminal book of sweeping exploration and discovery."

Duane Elgin, author of *The Living Universe* and *Voluntary Simplicity*

"A compelling tour de force of interdisciplinary scientific scholarship . . . a revolutionary mind expanding and soul restorative plunge through the rabbit hole . . . a must-read for anyone even vaguely attuned to the importance of evolutionary human spiritualization—highly recommended!"

Oscar Miro-Quesada, founder of The Heart of the Healer (THOTH), transpersonal psychologist, and originator of the Pachakuti Mesa Tradition of cross-cultural shamanism

"This brilliant book will, I believe, serve as a major catalyst to the most profound evolutionary moment facing humanity today—the recognition of who we truly are. . . . If you are questioning your place and purpose in the Cosmos, *The Cosmic Hologram* will lead you to your true identity."

Diane Williams, founder of The Source of Synergy Foundation

THE COSMIC HOLOGRAM

In-formation at the Center of Creation

JUDE CURRIVAN, Ph.D.

Inner Traditions
Rochester, Vermont • Toronto, Canada

Inner Traditions
One Park Street
Rochester, Vermont 05767
www.InnerTraditions.com

Library of Congress Cataloging-in-Publication Data

Names: Currivan, Jude, author.
Title: The cosmic hologram : in-formation at the center of creation / Jude Currivan, Ph. D.
Description: Rochester, Vermont : Inner Traditions, [2017] | Includes bibliographical references and index.
Identifiers: LCCN 2016031300 (print) | LCCN 2016034564 (e-book) | ISBN 9781620556603 (pbk.) | ISBN 9781620556610 (e-book)
Subjects: LCSH: Cosmology—Miscellanea. | Science and spiritualism.
Classification: LCC BD511 .C87 2017 (print) | LCC BD511 (e-book) | DDC 113—dc23
LC record available at https://lccn.loc.gov/2016031300

Printed and bound in the United States

10 9 8 7 6 5 4 3

Text design and layout by Debbie Glogover
This book was typeset in Garamond Premier Pro, with Gill Sans, Raleway, and Frutiger used as display typefaces

Cover images courtesy of iStock Photo

To send correspondence to the author of this book, mail a first-class letter to the author c/o Inner Traditions • Bear & Company, One Park Street, Rochester, VT 05767, and we will forward the communication, or contact the author directly at **www.judecurrivan.com**.

For everyone who wonders
not only how *our Universe is as it is*
but why, *and for those who are willing to follow*
the evidence—wherever it leads

Contents

PART 2

◇◇◇◇◇◇◇◇

Our In-formed and Holographic Universe

PART 3

◇◇◇◇◇◇◇◇

Co-creating in the Cosmic Hologram

Foreword

This book is a tour de force. It asks, if you were to make a universe, of what would you make it? And how would you put together what you need in order to make it? These are questions one asks when one bakes a cake. They are questions of "what" and "how" that apply, however, to all things, including the universe that is the ensemble of all things.

But reading this book is far more than engaging in an intellectual pastime. It is serious: full of information about what there is in the world, and how what there is, is related together. It is one of the most information-rich books I have ever read. Reading it one feels like having been handed the recipe of a master chef for a perfect cake. Only this recipe is for making a perfect universe. There could not be a more ambitious undertaking than trying to find the formula for this recipe.

Yet it turns out that there is a still more ambitious undertaking, and this book addresses it. It is not only to find *what* the universe is, and *how* it is put together, but also *why* it is. This "why" applies also to our being in the universe; it asks about the meaning and purpose of our existence.

We learn that the question "why" does not call for recourse to a transcendental entity or an ad hoc assumption about the nature of reality. It can be posed, and a reasonable answer to it can be found, within the enlarged confines of the science now emerging at the new frontiers.

This book gives me a distinct "aha experience." While I have known many of the facts and theories it discusses, here I have rediscovered them in a new and convincing light. This is the aha experience par excellence: it illuminates not only my understanding of this or that thing but also my grasp of the fundamental nature of all things and the identity of the master chef who makes them.

Kosmos, in the original Greek sense, means "ordered whole." The deeper perspective offered in this book is of the cosmos as an integral, inherently interrelated whole, with its premises rooted in the discoveries that have come to light at the cutting-edge of twenty-first-century science.

The Cosmic Hologram reveals the information that has made a perfect universe—*our* Universe. It will stimulate the reader to taste the entire coherent cosmic cake Jude Currivan offers for our palate. I wish the reader *bon appétit.* This is one feast that every open and intelligent reader will always remember and recognize with relish.

Ervin Laszlo

Ervin Laszlo, Ph.D. is a systems scientist, integral theorist, and classical pianist. Twice nominated for the Nobel Peace Prize, he has authored more than 75 books and published over 400 articles and research papers. He is the founder and president of the international think tank the Club of Budapest as well as of the Ervin Laszlo Center for Advanced Study. He lives in Tuscany.

PREFACE

Indra's Net

Imagine a shimmering net of light, without beginning or end. At each node of its weave and weft sits a shimmering jewel, and these myriad multifaceted jewels reflect and are each reflected by every other in rainbow-hued radiance of ever-changing illumination. Its infinite oneness manifests in the uncountable creative gems through which its eternal evolution is continually inspired and co-created.

Some three thousand years ago or more, this ancient numinous image of the Cosmos was first expressed in the sacred Indian text of the *Atharvaveda* and termed Indra's net; it was the means by which the Vedic deity Indra, god of the heavens, created the appearance of the whole world. Now, its revelation of integral reality and self-reflection at all scales of existence is being rediscovered and restated in a less poetic but equally majestic and scientifically based language.

While this twenty-first-century revolution is being led by cutting-edge science, its empowering implications will profoundly affect all of us. For it's about to transform not only what we thought we knew about the physical Universe but also our perception of ourselves and the nature of reality itself.

I've been seeking to understand what reality *really* is since childhood and have felt impelled to be on a lifelong journey to ask not only how but also why the Cosmos is as it is. My scientific pursuit of answers began when I was about five years old. My already growing fascination in astronomy was rewarded that Christmas with the gift from my parents of a copy of *The Boys'* [!!!] *Book of Space* by British astronomer Patrick Moore.

A few years later, the quantum world also grabbed my attention, eventually leading me by my late teens to study for a master's degree in physics at Oxford University in the early 1970s. There I specialized in quantum physics, exploring the physical world at its smallest scales and the cosmological physics of relativity theory. I sought to understand the Universe both in its totality and at its most extreme conditions; it was an exciting time—a time not long after the Universe's big bang origin had been confirmed and abuzz with the newly discovered phenomenon of black holes. Above all, I was attempting to ground and bridge my own expanding perspective in a scientific understanding of our physical Universe.

Even then, however, I already appreciated the fundamental incompatibility of quantum and relativity theories, theories whose views of space and time are radically different and which, by the time I was studying at Oxford, more than half a century of scientific investigation had still failed to combine in a unified theory.

In my second year of study I shared my hopes that this impasse might be resolved with one of my teachers, Dennis Sciama. Recently arrived at Oxford from the University of Cambridge, Dennis kindly invited me to take part in a forthcoming discussion of the issues of black holes and the presumed so-called singularities at their centers that would be led by two pioneering researchers. As it was a postgraduate seminar, I was probably the youngest person there when Stephen Hawking, already stricken with motor-neuron disease, and his colleague Roger Penrose described how gravitational collapse of massive stars would theoretically lead to the existence of such space-time singularities.

Now both world-famous scientists, even back then, their brilliance was being recognized and both were about to become fellows of the prestigious Royal Society, whose president had once been Isaac Newton.

Inspired by the seminar and encouraged by Dennis, I wrote an essay on black holes and the emerging views of how their behavior might offer insights into a theory of quantum gravity (which aims to reconcile quantum and relativity theories by finding a way to quantize gravity) and entered it for a university prize. I was delighted to win the prize and, as a perennially impoverished student, the prize money of a then-generous twenty-five pounds. I little appreciated though how true my concluding words of the essay would turn out to be: "Our knowledge of the behavior of matter under such extreme conditions is so limited at the moment that the formation of black holes and singularities may prove to be the least of our troubles."

For, more than forty years later and despite the brilliance of scientists including Hawking, Penrose, and many others, the discoveries of so-called dark matter and dark energy—whose nature is as yet unknown—have caused scientists to recognize that their cosmological worldview, even as described by the continuing development of still conflicting fundamental theories, is now only able to describe some 5 percent of our Universe. The rest, as currently understood, is essentially "missing."

For me, however, even more important than all of this as yet unresolved incompatibility and lack of explanation (or at least how they have been interpreted by mainstream science) is how to understand, account for, and incorporate the nature of consciousness.

From the beginning of my lifelong quest to comprehend the deepest nature of reality, I've also been fascinated with the ancient wisdom of cultures such as Egypt and Vedic India. Both of these traditions sought to explain the world and the perception of reality in terms that I was coming to recognize in my own explorations. Their cosmologies deemed consciousness and an underlying cosmic intelligence to be the foundation of the Cosmos and primary to its expression in physical form, viewing this intelligence as all-pervading and essentially *being* all that we call reality.

As such, these perspectives then sought not only to explain the mystery of how the appearance of our Universe arises from such deeper reality but also to understand the meaning and purpose of all life.

It's taken until now for science to start to catch up with the metaphysical insights and experiences of sages, shamans, and seers throughout the ages that are described by the metaphor of Indra's net. Compelling in this regard is the hypothesis of the holographic principle, first mooted by Dutch theoretical physicist Gerard 't Hooft, who in 1993 put forward the proposal that all the information contained in a region of apparent three-dimensional space can be represented as a hologram of the information held on its two-dimensional boundary.[1]

As we go on, we'll explore and scrutinize the unfolding evidence that our Universe is indeed a cosmic hologram, embodying its innate attributes of self-similar patterns of information and harmonic order that underlie all physical appearance on all scales of existence.

Having followed the progress of this emerging holographic perspective for more than twenty years, it seems we finally have a perception of reality that truly has the potential to offer a "theory of everything," a transformational and scientifically based model that recognizes that in-formation, consciousness, and ultimately cosmic intelligence are the ground states and all-pervasive fundaments of the whole world.

This profoundly empowering new understanding not only consolidates and builds on the discoveries and insights of twentieth-century science but also expands far beyond them.

For to understand the holographic principle, the pioneering science of the twenty-first century is also beginning to comprehend the fact that information is indeed more fundamental than matter, energy, space, and time. As we'll see, the cosmic hologram is being revealed across many different fields of scientific research, from the tiniest physical level of the Planck scale, vastly smaller than that of the quantum, to the largest scale of our entire Universe and at every level in between—including the reality of our everyday lives.

We'll see how interweaving concepts including quantum information, emergent and evolutionary complexity, the holographic principle, and fractal geometries and entropic processes, progressively backed up with new discoveries and experimental evidence, are revealing that the entire awareness of the physical world arises from a deeper level of informational reality.

A comprehensive understanding is still emerging, and it will take more time and more discoveries for science to fully confirm and appreciate this incredible new vision of the cosmic hologram. However, even at this pioneering stage, its potential to revolutionize our perception of reality and our place in the Cosmos is, in my view, far too important to leave to scientists alone.

The cosmic hologram is the twenty-first-century equivalent of Indra's net, showing—rather like my mum's wonderful chocolate cake—how all the requisite information in the form of instructions, conditions, ingredients, a recipe, and a container have been present from the very beginning of space and time to make our "perfect" Universe, one in which ever greater complexity can evolve to the point of self-aware individuals able and curious to understand and explore the deeper nature of reality and their place and purpose in the Cosmos (and enjoy chocolate cake).

We'll also examine the enigma of light as the bridge that weaves together the manifested physical world and whose extraordinary characteristics are exquisitely structured to enable the maximum creativity of the intelligence of what Einstein termed cosmic mind to express itself and evolve within our Universe. And it is *our* Universe, one of many universes within a multiverse of an infinite and eternal cosmic plenum, the one in which we have evolved as self-aware beings.

Throughout, I hope you'll be as astonished and delighted as I always have been by the wonderful way, as Einstein once said, that our perfect Universe is enabled to manifest and evolve its incredible complexity from a set of fundamental principles that is "as simple as it can be but no simpler." In fact we'll use his key insight as a continuing way-shower to the integrity of our quest.

Above all, we'll investigate the increasingly powerful evidence that the perceptions of our minds, hearts, and inner and outer purpose are microcosmic jewels of Indra's net through which the appearance of the whole world of the cosmic hologram is co-created, experienced, and explored.

PART 1

How to Make a Perfect Universe

1
In-formation

What is conveyed or represented by a particular arrangement or sequence of things . . .

It from bit . . . Everything is information.

John Archibald Wheeler, Physicist

It from qubit.

David Deutsch, Quantum Information Physicist,
Centre for Quantum Computation,
Oxford University

This book could not have been written until now, as we didn't have the information. It is literally a time whose idea has come.

As we begin to understand how emerging twenty-first-century science is progressively describing physical reality as a cosmic hologram, we first need to appreciate the primary nature of information.

The laws of motion and thermodynamics that define how matter and energy move and how they interact are basically laws of information. The concept of information content and flow is starting to be used powerfully to describe physical phenomena at deeper and more all-encompassing levels than hitherto.

The two twentieth-century pillars of science, the quantum and relativity theories, are also being reevaluated as informational theories, a development that is being seen as having the potential to finally bring together these as yet unreconciled perspectives of our Universe.

Yet as we'll see, this is just the first step to a much more encompassing perception, one that not only aims to understand the completeness of the physical world but also proposes a cosmology that encompasses *all* aspects of existence and experience and seeks answers to the deeper question of not only how reality is as it is but also why.

We'll start by seeing how information is becoming viewed as being more fundamental than energy and matter, and indeed, space and time. Along the way we'll discover how the minutest Planck scale of our physical world, trillions of trillions times smaller than the quantum realm, is key to our developing insights regarding the primacy of information. And we'll appreciate that information really *is* physical and begin our journey to understand how it literally in-forms our Universe—while at the same time transforming our view of what we actually mean by the term *physical*.

AN INCOMPLETE REVOLUTION

About a century ago, all our notions about energy and matter, space and time, changed enormously. Until then, since the time of Isaac Newton and other pioneers of science in the seventeenth century, energy was seen as an attribute of matter and motion, and space and time were viewed as being absolute in themselves, essentially separate from each other and a passive background to the "real" stuff going on.

In the late nineteenth century, however, the cracks created by baffling phenomena—such as the inability to account for the energy radiated by a hot oven—that could not be explained by the prevailing theories of the time appeared and led to the emergence of two revolutionary new approaches in physics: relativity theory and quantum mechanics.

These theories showed that Newtonian physics wasn't wrong. In fact, we still use its principles in rocket science as well as for a multitude of everyday purposes. It's just that this classical physics is incomplete. The revolutionary breakthroughs brought about by relativity theory and quantum mechanics expanded upon Newtonian mechanistic principles, revealing that these latter were an approximation at our ordinary experienced level of a more comprehensive but radically different understanding of our Universe: a universe where energy and matter are complementary expressions of each other and where dynamic fields of influence replace the notion of apparently separate objects and hitherto unexplained "action at a distance." And furthermore, space and time are each relative to the position of an observer, and we can only consider them as invariant in their combination as a four-dimensional concept of space-time.

But just as Newton's insights, three centuries before, were only partial, so the scientific revolution of the twentieth century was too. For the quantum theory that describes the physical world at tiny scales appears basically incompatible with the theory of relativity that describes it at massive scales. Fundamentally, quantum theory has no notion of time, and the space-time and gravity of relativity theory aren't quantized. In attempts to resolve their differences, much research has focused on developing a quantum description of gravity, with clues emerging from the study of matter under the extreme conditions at the beginning of our Universe and in black holes. Yet despite some eighty years of trying, the two theories still remain stubbornly unreconciled.

As if this wasn't enough, astronomical measurements of unexpectedly high speeds of galactic rotation, first convincingly reported by Vera Rubin in the 1960s, showed that something that has become known as dark matter, only measurable by its gravitational effects, is needed to keep the visible stars in their orbits and prevent them from being flung outward. Since then dark matter, most commonly proposed as being an unknown, weakly interacting but massive, subatomic particle, or WIMP, has been shown to pervade our Universe.

Beginning in the 1990s, measurement of the rate of expansion of space has also revealed that, instead of slowing down as expected, it's actually accelerating[1]—a situation currently attributed to the existence of so-called dark energy. While there are a number of candidates for this cosmological constant that appears to be an inherent tension in the fabric of space-time itself, none are as yet proven.

Neither of these two "dark" and utterly unanticipated components of the Universe are as yet understood, but in constituting a currently estimated 95 percent of all its energy and matter, they relegate the two foundational, but as yet unresolved, theories of twentieth-century science to describing a mere 5 percent of physical reality.

So where do we go from here?

As was the case in previous scientific revolutions, it's in seeking to deal with such apparent differences, inconvenient anomalies, and exceptional phenomena that will open the door to our greater understanding and an ever more comprehensive view of the world.

It may, however, well prove to be the case that dark matter and dark energy will be able to be incorporated within the expanded bounds of twentieth-century physics. For me, it's the quest not only to reconcile them but also to respond to the deeper question of *why* quantum and relativity theories appear so adamantly opposed to each other that may prove to be instrumental in profoundly revolutionizing our understanding of the physical world, as well as our understanding of consciousness and the nature of reality itself.

Even though the technological potential of these two theories has been exploited, other questions, such as why and how the act of observation renders a quantum entity "real" and what nonlocal connectivity—which Einstein called "spooky action at a distance"—*really* means, have largely been relegated to the sidelines of research.

It's only now that deeper clues, while they are still embedded within quantum and relativistic phenomena, are being considered and rooted in a far more wide-ranging and much deeper perception of how the whole world is co-created at all scales of existence. I think Einstein would be pleased. This emergent understanding of informational physics, however, while based on leading-edge scientific discoveries, represents a radical challenge to mainstream science. Thankfully though, it's the power of the scientific method and its practical and unbiased approach that will compel mainstream science to follow the experimental and experiential evidence, regardless of an entrenched worldview or the theories that attempt to make sense of it.

Accumulating evidence from discoveries and insights in the past few years across a far-reaching range of disciplines allied with a dramatically different approach to theoretical paradigms is progressively describing our Universe in terms of holographic information and ushering in a new twenty-first-century scientific revolution. The potential of this radical new perspective is far greater than its ability to profoundly transform our perception of the physical world once again. As we'll go on to explore, because it's realizing that nonphysical and multidimensional information is required to understand our Universe, this emergent vision will bring about perhaps a more fundamental revolution in our comprehension of ourselves and the whole world than ever before.

The progressive recognition that scientific laws can be redefined and expanded in terms of information goes back to the nineteenth-century study of thermodynamics. This is the branch of physics that investigates heat and temperature and how they relate to energy, work, and entropy. Previously, thermodynamics was most commonly understood as the measure of order or disorder in a system, but it is now being viewed more fundamentally in terms of its information content and flow.

From his study of the behavior of gases in the late nineteenth century, Austrian physicist and philosopher Ludwig Boltzmann predicted the existence of atoms and molecules. He was opposed by the majority of physicists at the time, and this opposition contributed to the deterioration of his already fragile mental state, leading him to commit suicide in 1906. His death was even more tragic as within a couple of years the reality of atoms was confirmed and his theories were vindicated.

In the middle of the twentieth century, in his study of communications—in other words, the transportation of information—Bell Labs scientist Claude Shannon showed that the mathematical formula that describes the entropy of the energy of a gas in thermodynamic terms and the information content of a system is *exactly* the same.

The very simple but hugely important equation that connects entropy with both energy and information should soon become as iconic as Einstein's famous equation linking energy and matter, and it certainly stands alongside it in importance. In this equation, $S = k \log W$, S stands for entropy, k is a constant named after Boltzmann, and $\log W$ is essentially the logarithm of the energy states or informational content of a system.

As we'll see, in the past few years this equivalence of information and energy and the relationship with the concept of entropy, the equation for which is inscribed on Boltzmann's memorial, is revealing that information is actually *more* encompassing and fundamental, indeed expressing itself in complementary ways through the conservation of energy and matter and entropically through the flow of time itself.

In 1666, with the University of Cambridge temporarily closed as a precaution against the ravages of the Great Plague then raging through the land, a young Isaac Newton, sheltering at his home in the tiny Lincolnshire hamlet of Woolsthorpe, used his time very well. In addition to his studies of optics and the development of calculus, the mathematical formulation of change, it was here that Newton really did see an apple fall from a tree in his garden, as he later told his biographer William Stukeley, and which led him to discover the law of gravity.

Some years later he was able to show mathematically that the rules of elliptical planetary motion, observed by Johannes Kepler earlier in the seventeenth century, obeyed the same law. By doing so, he proved that gravity not only pertains on Earth but also that it governs our entire solar system and so, by implication, pervades our entire Universe.

Indeed, all fundamental laws of physics are *universal* in that they apply throughout our Universe, regardless of circumstances of space or time. Such universality seems so obvious to us that we tend to overlook its enormous significance in revealing the all-pervasive and intrinsic interconnectivity of our entire Universe and all the phenomena that manifest within it.

No subsystem, whether an elementary particle, a person, a planet, or a galactic cluster within our Universe, is or can be completely isolated. Everything at all scales of existence is being progressively discovered to be inherently related by in-formational content, flows, and processes as we'll see as we investigate further, this isn't just basic data but all-pervasive in-formational patterns and relationships.

We often use the word *random* to describe a lack of configuration or order. Yet, as theoretical physicist and philosopher David Bohm is reported to have said in a 1987 interview with holistic researcher and author David Peat, "Randomness is assumed to be a fundamental but inexplicable and unanalyzable feature of nature, and indeed ultimately of all existence. . . . However, what is randomness in one context may reveal itself as simple orders of necessity in another broader context."

While they may indeed be probabilistic, phenomena at the quantum level (and indeed at all scales within our Universe), they aren't "random" or based on chance, as is often portrayed. They act according to the probabilities within a range of possibilities that are dependent on the information they embody.

Indeed, as the emerging view of the cosmic hologram is revealing, nothing in our Universe is ultimately random; everything that manifests in the physical world emerges from deeper and ordered levels of nonphysical and in-formed reality.

INFORMATION *IS* PHYSICAL

A couple of years ago when I fell down a flight of stairs and slammed into a brick wall at the bottom, badly breaking my arm, my entire body screamed in pain, and the physical world felt very, well, physical! It wasn't the moment to remind myself that all that we term physical is essentially 99.999999999999 percent no-thingness, and the rest is excitation patterns of energy and information.

This fact doesn't just reflect the vast distances between planets, stars, and galaxies but the innate spaciousness of the quantum-scale realm. The familiar metaphor of a hydrogen atom, where if the single proton of its nucleus is represented by a basketball, and its single orbiting electron is then around three kilometers (1.86 miles) away, offers a visual image of this lack of materiality.

Even more insubstantial, atomic nuclei actually comprise fields of influence within which quarks, the fundamental entities that make up the nucleonic protons and neutrons, are deemed, according to the current standard model of quantum field theory, to be point-particles. In other words, they have no internal composition or essentially spatial extent, or they have one that's too small to be measurable by current techniques. This too is the case for electrons and indeed all so-called elementary particles that are viewed as being the most basic constituents of energy and matter.

With such a transitory basis for our notions of what's physical, it's only the so-called Pauli exclusion principle, named after the Danish physicist Wolfgang Pauli, that prevents the atoms of our bodies from coalescing with the matter of our surroundings.

This principle states that no two quarks or electrons can occupy the same quantum state—that is, being in the same place at the same time with the same quantum attributes. This rule not only governs the chemical behavior of atoms, giving rise to the elements of the periodic table and their great diversity of properties, but also, in preventing the subatomic components of atoms from approaching each other too closely, accounts for the essential stability of matter.

In the past few years the emergence of M-theory, the successor to multiple string theories of the 1980s, aims to reconcile quantum mechanics and gravity, replacing the zero-dimensional point-like particles of the standard model with vibrating one-dimensional strings (which in M-theory are associated with holographic boundaries called *branes*). These too, however, are deemed to have no internal structure. Although M-theory (first conjectured and its name coined by Edward Witten in 1995) and other competing theories are offering progressively deeper insights, all of them merely replace one ephemeral basis of fundamental physicality with other similarly ghostly alternatives.

As we'll see later, only by our considering the will-o'-the-wisp nature of physicality in terms of the more fundamental nature of information are we able to perceive the primordial realities from which our experience of physical reality arises.

In 1931, Max Planck, one of the greatest pioneers of quantum theory, stated that "I regard consciousness as fundamental. I regard matter as derivative from consciousness." Many other scientists before and since, including Einstein, have expressed similar views.

In 1991 information theorist Rolf Landauer was the first to explicitly declare the physical and fundamental nature of information[2] that is itself inherent in all notions of consciousness. We'll explore this crucial insight from three main angles: how quantum behavior can be explained in terms of information; how information and the fundamental physical concept of entropy are vitally interconnected; and how all physical laws and natural phenomena can be re-expressed informationally. In all three approaches we'll see how all universal physicality can be considered and restated as processes, states, and flows of information.

Further evidence that information is an intrinsic attribute of all physical systems was first suggested by Hungarian American physicist Leo Szilard, who theoretically showed that there is a minimum amount of work needed to store one piece of digital information, or bit.[3] Landauer then showed that the erasure of any such bit of information increases entropy by precisely the same amount: $kT\ln 2$, where k is Boltzmann's constant, T is temperature, and $\ln 2$ is the natural logarithm of 2—a rule now known as Landauer's principle.

Finally, in 2012, physicists Antoine Bérut, Eric Lutz, and their colleagues were able to experimentally prove Szilard and Landauer's predictions by measuring the dissipation of heat from the erasure of one bit of information.[4] Reporting in the journal *Nature,* they succeeded in verifying the link between the role of heat and temperature in the relationship between energy and information and so confirmed the essential physicality of the latter.

QUANTUM INFORMATION

Considering information as fundamental offers us a deeper understanding of why matter and energy are quantized and also of the wave-particle complementary nature of quantum entities. However, before we move on to considering quantum theory in informational terms, let's begin by briefly reviewing these two key cornerstones in energetic terms as its pioneers understood them.

Before their insights, classical physics considered our Universe as a continuum. It was by attempting to explain something apparently as simple as how a hot oven radiates heat that led Planck to propose the first theoretical quantum perceptions.

The problem was that the classical electromagnetic theory of heat and light maintained that such radiation, from its most idealized source in a so-called black-body, should occur on a continuous and unlimited basis. This conclusion resulted in the total radiated power being theoretically infinite, which was a nonsensical result and one that the bemused physicists of the time referred to as the ultraviolet catastrophe.

It was Einstein though, taking and applying Planck's insights, who solved the problem. He was awarded the Nobel Prize in 1922 for this solution rather than for his more famous discoveries relating to the equivalence of energy and matter and the relativity of space and time.

Einstein's explanation came about through his studying of a problem related to the ultraviolet catastrophe, that of the so-called photoelectric effect, where incoming light can dislodge electrons from the surface of a metal plate. In classical theory, where light is perceived as being continuous, electrons should be able to be freed by shining more intense light onto the plate or doing so for a longer time. However, this isn't what is observed.

By realizing that only light above a certain energetic threshold of frequency is able to free the electrons, regardless of the intensity or duration of the light beam, Einstein postulated that light travels in discrete packets, or Planck's theoretical quanta, whose energies increase with frequency.

So as a light beam of progressively higher and so more energetic frequency (say, from red to blue and then ultraviolet light) is shone onto the plate—even if at very low intensity—individual quanta have the requisite energy to bump electrons out of the metal.

Einstein's genius was in recognizing that the photoelectric effect as a result of quantization also explains the conundrum that Planck himself was attempting to understand: the finite radiation of a hot oven. Here too emitted energies are expressed, not as the classical continuous and unlimited spectrum of wavelengths, but as discrete quanta each having a specific energy and so totaling a finite amount.

Quantization of energy and matter, however, is only one side of the complementary relationships between waves and particles that lie at the core of quantum theory. While taking different stances on what it actually meant in terms of the nature of reality, in the early twentieth century pioneering physicists such as Niels Bohr and Werner Heisenberg as well as Einstein and Planck recognized that physicalized matter and energy could be regarded as both waves and particles: for example, the wave-like energies of electromagnetic waves of light also behave as quantized particular photons. In 1924 the French physicist Louis de Broglie then took the next logical step and realized that all elementary particles also exhibit wave-like behavior, albeit at very small wavelengths.

It was in 1926 that, in a series of scientific papers published in an intense period of only a few weeks, Austrian physicist Erwin Schrödinger produced the equation named after him that has become foundational for understanding how the mechanics of quantum entities behave as well as their wave-particle complementarity.

Schrödinger's deceptively simple equation is made up of a series of variables within a single wave function that incorporates *all* the possible quantum states of a system in terms of probabilities, which are spread throughout the waveform. The equation's most general form shows how the relative probabilities of these states evolve through time. Only when observed or measured do the probabilities then become a specific state in the physical realm.

Rightly seen as being one of the most important scientific insights of the twentieth century and honored with a Nobel Prize, Schrödinger's equation reveals the most complete amount of information that it is possible to know about a physical system, not only at quantum scales but also at every level up to and including our entire Universe.

Schrödinger's wave function, though, predicts the likely physical manifestation of a system when observed in terms of the underlying probabilities of how it evolves over time, not in the physical realm but in the so-called complex plane. For many years this was seen as a mathematical abstraction that happens to be very useful in predicting quantum behavior. But more recently, as we'll see in further detail later, the complex plane and indeed other nonphysical spaces and dimensions are being ever more recognized as a deeper level of reality from which the physical appearance of our Universe emerges.

Having briefly described the quantization of both matter and energy and their complementary wave-particle attributes, we can now start to reframe them in terms of information.

Fundamentally, as the need to resolve the ultraviolet catastrophe of classical physics showed, a continuum has the capability of literally carrying infinite information. Our Universe, however, is finite. Within it, time literally began some 13.8 billion years ago at the big bang. And, given our understanding that physical space and time can only be considered as a combined entity of space-time, finite time must correspond with finite space.

A finite universe can only embody finite information. There thus needs to be a mechanism for the essentially unlimited potential of the nonphysical wave function to become finitely manifest. Quantization, with its innately discrete nature, is such a mechanism, enabling finite information to be expressed within space-time.

We're familiar with the idea of the ones and zeros of bits being the digital building blocks for computerized data processing. But such bits are the simplest expression, the literal building blocks, of all finite information, enabling its processing with the minimum energy and maximum stability. Combining bits, rather like building a house out of bricks, enables any and every possible outcome—and multiple outcomes—to be expressed as efficiently as possible.

Essentially the energy and matter of our Universe is quantized because information is digital and thus quantized, and information is quantized because it is the most effective means for its communication.

REAL-IZATION

From the viewpoint of information, we can also better appreciate how wave-particle complementarity enables energy-matter to be observed in one or another form. The key is in how the definite state of the entity is observed and measured—in other words, in *how* information about it is accessed. Essentially the realization derives from the information interaction between the entity and its environment. Until that moment the Schrödinger equation describes the evolving probabilities and potentialities of both aspects of its wave-particle nature.

This understanding of how real-ization comes about has been tested by ever more sophisticated means over the past few years. A preliminary and key question was answered in 2012 in a groundbreaking paper that reported the findings of a so-called quantum delayed-choice experiment. This experiment aimed to resolve whether quantum entities behave either as a wave or a particle depending on the circumstances, or whether they are *always both until they're measured;* when dependent on the type of measurement, they switch to one form or the other.

For the first time, the research team was able to construct a measuring process that included a high level of nonlocal behavior, enabling and revealing photons to exist in a simultaneous state of both wave and particle, as lead author of the paper, Alberto Peruzzo of the University of Bristol in the UK, confirmed.[5] Their nonlocal entanglement enabled the photons in the experiment to delay the choice of whether to behave as particles or waves. The nonlocality effectively replaced the delayed choice of an observer, which would have triggered either their wave-like or particle-like form to be exhibited.

Knowing that the quantum choice to prompt either its wave-like or particle-like aspect does indeed arise from observation, scientists have still sought to determine why the quantum entity sometimes then behaves as a wave while at other times as a particle. And still, scientists being scientists, they've tried to see if there's some way of observing both at once.

In 2004, and then with a more refined experiment in 2006, Iranian American physicist Shahriar Afshar baffled his colleagues when he appeared to violate the either-wave-or-particle behavior by seemingly demonstrating wave-like interference patterns of light while at the same time measuring the paths taken by the quantum particles of light: photons.[6] In a subsequent experiment in 2012, a team of German scientists also set up an experiment that seemed to verify Afshar's findings.[7]

In 2014 though, a collaboration of Canadian and US physicists including Eliot Bolduc and Robert Boyd reanalyzed the challenge posed by these apparent findings and discovered a flaw that comes back to how information is accessed.[8]

They realized that the setup for any experiment must choose how it obtains information through observation and measurement—that is, how the experimental apparatus interacts with its environment. For example, a choice can be made to measure the experiment-environment aspect with the best information about the path or, alternatively, to choose that with the best visibility of the interference fringes of wave-like behavior.

To properly test complementarity, the measurements must be equally sensitive to all the possible states of the system, a requirement known as fair sampling. When this requirement was applied to the German team's 2012 experiment and Afshar's earlier analyses, the 2014 team determined that they violated the fair sampling rule. It's now incumbent on any future testing of wave-particle complementarity to ensure that such fair and unbiased measurements of information are rigidly enforced.

The 2014 findings confirm the evidence of all past experiments that point to wave-particle complementarity being an intrinsic attribute of space-time. The progressive understanding though of how information underlies and pervades the appearance of such complementarity is crucial to realizing how the environment and observer are integrally interconnected with any and all experiments. In other words, there's no separate "objective" reality, and the entirety of our Universe is an integrated, coherent, and in-formational entity.

There is literally no "environment" apart from the consciousness of the "observer."

✺

Before we go on, and at this point in our investigation, let's briefly consider how, prior to the emergence of the informational approach and holographic perspective, quantum physicists had come up with a hypothesis that I think will be seen in the future as one of the most unhelpful diversions in the history of science.

This hypothesis is called the many worlds interpretation, or MWI, and is one of a number of versions of the concept of a multiverse. The MWI describes the nature of reality as consisting of infinite *parallel* worlds, embodying infinite alternative histories, presents, and futures and maintains that all are equally "real."

The MWI was a contrasting response to the interpretation of experimental evidence that the observation of an event affects its outcome: the so-called Copenhagen interpretation. As we'll see, more and more evidence is accruing that not only supports the Copenhagen perspective but also expands it to the perception of an informationally based, integral, nonlocally interconnected and ultimately intelligent Universe.

However, in 1957, and before this twenty-first-century perspective gained traction, American physicist Hugh Everett proposed the scenario of the MWI.[9] In the 1960s and 1970s it was popularized by fellow US physicist Bryce Seligman Dewitt and entered the speculations of science fiction.

The MWI seeks to avoid the underlying and all-pervasive presence of information, intelligence, and consciousness in manifesting reality by the premise that *every* possible outcome of *every* quantum-scale event really exists by branching off somehow into an infinity of other universes. Essentially, it aims to banish any deeper consideration and involvement of consciousness from understanding the nature of reality.

Not only is there no mechanism for such a scheme, but it also violates pretty much every principle of physics and adds a level of inexplicable and unnecessary complication without any rhyme or sensible reason. (Now that I've got that off my chest, let's move on.)

Just as any quantum system exists as an evolving state of superposition of probabilities until its real-ization, so too does the information of its possibilities exist as superposed so-called qubits (quantum bits) until real-ized as specific digitized bits. This enables us to expand the understanding of information to quantum information that integrates with quantum theory and expands it beyond being a description of energy and matter but being more fundamentally expressed in terms of information.

Qubits are the informational building blocks embedded within the probabilities of the Schrödinger wave function, just as digitized bits are then the information specified by the "it" of its real-ized single state.

Unlike bits that can only exist as either zero or one (or any dualistic on/off expression), qubits can be superposed in any combination of the two states. Until they interact, are observed, or measured in some way, there's no way to access any information about qubits. However, once they devolve to a specified state, and are then described by bits, there's also then no way to retrospectively access information about their superposed states prior to such measurement.

These attributes of qubits have spurred accelerating research into quantum computers. Their ability to exist simultaneously in many states of superposition enables them to process information at vastly greater speeds than a digitally based computer, and the inaccessible nature of such superposed states can be used to create effectively unbreakable cryptographic codes.

We'll shortly continue to reframe the phenomena of our Universe in informational terms, but to do that we first need to begin our exploration of the minutest and most extreme scales of energy and matter, space and time.

THE PLANCK SCALE

The primary laws of our Universe incorporate a number of unchanging physical constants. Three of these universal factors govern the relationships between energy-matter, space and time. These are the speed of light, the gravitational constant, and Planck's constant. Respectively, each is associated with: special relativity, gravity, and quantum mechanics.

As Planck was the first to suggest, these constants can be combined to define a scale of measurement that is integral to the workings of our

Universe. This scale, arising naturally from the interplay of universal forces, appears literally to represent the possible limits of physical reality itself.

The extent of its measure of space (or Planck length) is an astoundingly tiny 10^{-35} meters, which is as small when compared to the size of a soil amoeba as such a tiny organism on Earth is to the size of our entire Universe.

The Planck scale of time, which denotes the time taken by light to travel a Planck length, is an almost equally unimaginable 10^{-44} seconds, which is even more extraordinary when it's compared to the age of our Universe since the big bang—around 10^{17} seconds.

The Planck scale of energy-matter denotes where quantum and gravitational forces become equivalent as in the extreme conditions at the beginning of the Universe and within black holes. The Planck mass of 10^{-8} kilogram is the maximum mass for a point-like elementary particle and equivalent to a black hole with a diameter of one Planck length.

The Planck scale, though, is far more than an interesting scientific curiosity. For, as we'll discover, investigations at these extremes, including the entropic informational content encoded by black holes, are not only revealing that the nature of space-time itself and the relativity theory that describes it are best understood in informational terms but also that the informational bits entropically embedded by space-time are pixelated at the Planck scale.

The complementary nature of wave-particle phenomena is also revealed by consequential restrictions on the ability to measure associated pairs of physical attributes beyond the Planck scale: the energy of an entity and the time of its measurement, and the momentum of an entity (its mass multiplied by its velocity) and its position in space. These limitations are wrapped up in a basic premise of quantum theory called the Heisenberg uncertainty principle.

As with so many terms coined by science, the so-called uncertainty principle is poorly named, for it is rather a principle relating to *indeterminacy*. Its deeper insight is of the innate complementary nature of certain universal physical attributes and their measurement. The more exact the one measure is, the less precise its complementary partner. There is a fundamental limit to the accuracy of one to the other that relates to the Planck scale.

This innate boundary has nothing to do with our ability to take specific measurements but is essentially caused by, and a further clue to, the pixelation of space-time at that minute scale.

IN-FORMING THE WHOLE WORLD

From the Planck scale upward to the entirety of our Universe, we're beginning to appreciate how everything we call physical reality is literally made up of information. Recent experiments have also shown that quantum informational behavior isn't limited to the quantum scale. Exactly the same types of wave-particle complementarity and superposition have been demonstrated for entities as large as organic molecules. They are so enormous when compared to the tiny quantum level that such evidence is revealing that there's actually no difference between the informational behavior of quantum entities and that of macroscopic ones.

The reason why larger-scale phenomena don't *appear* to exhibit quantum-scale behavior is that it's progressively more difficult to isolate them and the information inherent in them from their interactions with their wider surroundings—surroundings that we incorrectly view as being "separate" from them and term the environment. For, as we've seen, it's accessing such information that manifests the phenomenon, any phenomenon at any scale, from a superposed probabilistic state to a specific state of real-ization.

Just as quantization and Planck-scale pixelation have perfect informational attributes for the manifestation of the physical world, so do holograms and the generic nature of the holographic principle.

Holography was invented, or more correctly, discovered, in 1947 by Hungarian British physicist Dennis Gabor.[10] While man-made holograms are rudimentary, as we'll see later, advances in holographic technologies are nonetheless offering deeper insights into the nature of the cosmic hologram. For now though, it's enough for us to note how holographic characteristics are ideal for the recording, storage, processing, and retrieval of information, both on a distributed basis and nonlocally.

The interference patterns of light that make up holographic projections are able to incorporate enormous and unmatched amounts of information. In addition, at its smallest coherent scales, every part of a hologram contains the informational content of the whole, effectively distributing and storing it nonlocally and self-replicating the macrocosmic whole within its microcosmic elements.

Bearing in mind what we've learned so far, let's now turn to examining the larger-scale attributes of our perfect Universe and begin to explore the informational foundation of space-time itself.

2
Instructions

Instructions tell something how to do something . . . .

This planet came with a set of instructions, but we seem to have misplaced them. Civilization needs a new operating system.

Paul Hawken, activist and author of *Blessed Unrest*

Just as for baking a perfect chocolate cake, making our perfect Universe, which after around 13.8 billion years of evolution has arrived at the point where we can enjoy cake, requires the essential initial instructions, conditions, ingredients, a recipe, and a container.

We'll now explore how the necessary instructions, the algorithmic programming that literally in-formed the creation of space-time and the energetic and informational principles that perpetuate throughout the lifetime of our Universe cycle and that are the foundation for its evolution, were present from its first moment.

OUR UNIVERSE

Until the 1920s, our Universe was deemed unchanging and infinite by most astronomers. However, in 1927, Georges Lemaitre, a Belgian Catholic priest and scientist, put forward the idea, confirmed observationally two years later by Edwin Hubble (after whom the Hubble telescope is named), that instead of being invariable the Universe is expanding.

Nonetheless, astronomers such as the UK's Fred Hoyle still viewed it as being ultimately infinite, coming up with the now-disregarded steady-state model that theorized that matter continually forms to offset the expansion and retain a constant and ultimately eternal state. Hoyle compared his proposal with the contrasting theory of a specific and thus finite beginning, for which in 1947 he coined the name big bang.

The big bang model, however, made a testable prediction: that from an enormously hot start, as our Universe expanded it cooled down and should accordingly be filled with remnant radiation from that early era that would now be measurable as very cold and radiating at microwave wavelengths. Only in 1965, when Bell Telephone Laboratory engineers Arno Penzias and Robert Wilson accidentally discovered this so-called cosmic microwave background, or CMB, when they were bothered by excess noise in the radio receiver they were building, was the prediction confirmed, which led to the big bang becoming generally accepted. Initially they thought pigeon poo in the equipment was to blame for the noise and went to great (and presumably messy) efforts to remove it. But when the noise persisted, other scientists realized that it was the residual energy signature dating back to around 380,000 years after the big bang, when our Universe first became transparent to light.

Not only the CMB's presence but also analysis by NASA's Wilkinson Microwave Anisotropy Probe (WMAP for short) of tiny irregularities within it provides further evidence that our Universe is finite. In 2003, WMAP's examination of the patterns of slightly hotter and colder spots that represent ripples in it, rather like minute waves in the ocean, showed that longer wavelengths were missing. An infinite universe should include wavelengths of all lengths, whereas a finite universe will have a cutoff, such as that measured.

While there are cosmologists who still maintain that despite its finite birth, our Universe will expand to infinite spatial extent and eternal temporal duration, as we'll continue to see, such a view is at odds not only with fundamental logic but also with the laws of physics and increasing observational evidence.

To conclude, however, that the space-time of *our* Universe is finite is not to presume that the entire Cosmos is, nor indeed that its infinity is only expressed at the physical level. In the past decade or so a plethora of theoretical models of other and indeed all possible universes forming a so-called multiverse have been proposed. These have attempted to address various contentious issues, including the apparently very specific and special nature of *our* Universe, and most usually by viewing it as being a random event within an infinity of other random events—all equally without meaning or purpose.

As we've already begun to perceive, to understand the physical realms not only of our Universe but also any multiverse scenario calls for a deeper perspective of the ordered information that underlies and pervades them. Toward the end of our exploration, we'll consider certain multiverse theories in informational terms too, to assess the extent to which they may offer us an expanded perception of an infinite and meaningful Cosmos.

However, in the meantime, we'll continue to focus on our Universe. We have an exciting journey ahead.

THE BIG BANG

The big bang wasn't big. Nor was it a bang.

At this first moment of our Universe, literally as space and time began, it was minute and embodied impeccable order. Then, rather than starting with a chaotic explosion, it expanded instead with amazingly exquisite precision.

From its birth its instructions encoded the information enabling the creation of elementary particles and the fundamental processes and interactions that progressively gave rise to stars, galaxies, and the emergence of ever more complexity.

The same ancient Vedic sages of India who envisioned Indra's net also described the beginning and evolution of our Universe in a similarly ordered and expansive way as an out-breath of the cosmic creator Brahma.

So let's now take a look at how extraordinarily special and exact the genesis of our Universe was and indeed what was required for us to be here billions of years later.

The initial level of order itself is astonishing. Analysis of the CMB in recent years has revealed only tiny energy irregularities of less than one part in a hundred thousand.

To appreciate the smallness of such variability, the Challenger Deep in the Marianas Trench of the Pacific Ocean was measured in 2010 as being about 11,000 meters in depth. If the cosmic ripples in the CMB were replicated as waves, they would be equivalently only ten centimeters high on the surface directly above this deepest point in Earth's oceans.

It turns out that both its very high level of initial order and yet this minuscule level of variance are both crucial to the evolution of our perfect Universe. However, there is a major difficulty for cosmologists in that this enormously high level of uniformity extends across the entire observable Universe. For in doing so such homogeneity covers distances far greater than what would have been able to be in contact by signals traveling at the speed of light, the cosmic speed limit within space-time. Unable to explain this extremely high level of self-similarity that spreads beyond the spatial horizon limited by the light speed, cosmologists have called it the horizon problem.

In addition, measurements of the total average energy and matter density of our Universe show that within 0.4 percent margin of error, a tiny amount, it's *exactly* the critical density that enables the geometry of our Universe to be flat. In other words, the entirety of space-time obeys the Euclidian geometry that we were taught in school. For such geometry, drawn on a flat sheet of paper, the three inner angles of a triangle add up to 180 degrees. This very specific geometry contrasts with the much greater number of options for non-flat geometries: for example, a triangle drawn on the so-called open surface of a concave saddle shape or on a closed surface such as a sphere where the variability of its combined inner angles are then generally less or greater than 180 degrees, respectively.

Such flatness, though, is crucial for the relativity of space and time and their invariant combination as space-time. In addition, it tells us about the future fate of our Universe: for the critical energy-matter density that gives rise to flatness is also exactly sufficient to eventually halt the expansion of space but insufficient for it to re-collapse.

Not only is flatness or something very close to it far less likely than the wide ranges of other geometries, but embodying the critical density for flatness as our Universe expands requires even more exceptional fine-tuning from the first moment of the big bang. Though to say "exceptional" is rather an understatement. For at that initial instant, space-time could not have departed from complete flatness by more than one part in 10^{62}. Cosmologists, flummoxed by such incredible exactitude, called this the flatness problem.

Given that homogeneity on a universal scale and precise flatness are actually intrinsic requirements for our perfect Universe to evolve as it has, what cosmologists mean by describing something as a "problem" is due rather to trying to fit it within the big bang scenario. Their dilemma has been to try to reconcile their view of a randomly generated universe with the incredible level of order and fine-tuning ours actually embodies, of which the flatness and horizon conundrums are but two examples.

We'll go on to explore an alternative rationale that not only offers a resolution to these two perplexing attributes but other apparent anomalies of the big bang origin of our Universe as well. But before we do, let's briefly discuss the proposed explanation for them presently adhered to by most cosmologists.

This is the view that our early Universe went through a so-called inflationary epoch. This idea of a period of exponential expansion, first put forward by theoretical physicists Alan Guth and Andrei Linde in the 1980s, proposes that a tiny moment after the big bang, space expanded at least a hundred trillion trillion times in about a thousandth of a second.

Although relativity theory prohibits any signal within space-time moving faster than the speed of light, there's nothing in the theory to prevent space *itself* expanding faster than the speed of light. The inflationary hypothesis maintains that such an incredibly rapid expansion enabled the universal self-similarity we observe. In addition, such a process is suggested to have smoothed out virtually all initial irregularities and flattened any previous curvature of space.

While incorporating amendments by others, seeking to overcome issues with the original proposal, there's as yet no convincing theoretical basis for such an inflationary mechanism, nor how and why it would apparently switch on and then off.

It's this last issue that is perhaps the most problematic, and virtually all the inflationary ideas so far put forward haven't been able to address it. Instead they project what's known as eternal inflation: in other words, a runaway inflationary process that just continues, but beyond our observational abilities that are limited by the speed of light from reaching us, and yet infinitely expanding our Universe. With the inflationary big bang model, we've essentially returned to a postulate not so very different from Hoyle's infinite steady-state theory, albeit it with a finite beginning.

Whereas Hoyle may have been gratified, as a famously blunt Yorkshireman, I suspect he, like me, would have balked in no uncertain terms at the conjoining of finite and infinite premises in this way. Nonetheless, by appearing to resolve the horizon and flatness "problems" of a universe that otherwise would require a paradigm shift in perspective, the inflationary hypothesis is at present widely accepted by cosmologists, and the search to discover its energetic signature goes on.

In March 2014, measurements by the BICEP2 (a helpful acronym for the mouth-filling Background Imaging of Cosmic Extragalactic Polarization) experiment sensationally reported a signal of polarized light from the CMB, which if confirmed would have offered powerful evidence for inflation. In early 2015 though, thanks to including the latest data from the Planck satellite probe, the BICEP2 team acknowledged that they were mistaken and that instead of arising from the

CMB their signal actually comes from polarized dust in our own Milky Way galaxy.

I mention their erroneous interpretation because it provides an example of how, sometimes, subjectivity infringes upon scientific objectivity. In wanting to confirm the premise of inflation, the researchers didn't wait until the preliminary evidence was validated; their subjectivity prevailed. While proponents of inflation continue to adhere to their proposition, it does impede a willingness by many cosmologists to look at other explanations.

Let's move on from the hypothesis of inflation and instead explore the evidence for an alternative viewpoint of how the instructions that in-form our Universe do so. To do this, we'll consider how two of the most important laws of physics help point us in the right direction, and especially when restated in informational terms.

THE FIRST LAW OF INFORMATION

For an isolated system, in other words, one that has no interaction with anything outside of it, the total amount of energy and matter within it remains constant, neither able to be created or destroyed but only to change form. This physical principle is known as the law of conservation of energy and is so primary that it's also known as the first law of thermodynamics.

Crucially, all observations from the quantum scale upward strongly support it being a universal law and so requires that our Universe in its entirety must be an isolated system.

At the quantum scale the Schrödinger wave equation, which as we've seen describes the evolution of a quantum system, shows that even if there is uncertainty in its specific manifestation, the *overall* probabilities of its various states don't change over time and so, in other words, its total energy is conserved.

At macroscales too, the principles of general relativity that describe space and time, and to which we'll soon return to explore more deeply, *require* universal conservation not only of energy and matter but also of momentum, the multiple of an object's mass and its velocity. What these and all evidence to date implies is that the totality of energy and matter present at the very first moment of space-time is *exactly* the same amount of energy and matter in our Universe today and will be *exactly* the same amount of energy and matter until its end.

This implies that *all* the energy and matter within space-time is conserved: not only visible energy and matter constituting, on current estimates, only some 5 percent of the total but also the 27 percent of dark matter and the 68 percent presently attributed to dark energy.

In addition to the total amount of energy-matter being conserved, the flatness of space-time, as we've seen, results from the average *density* of energy-matter of our Universe, over its lifetime, as being a critical amount that will ultimately bring the expansion to a halt.

The gravitationally attractive energies of dark matter have the same effect as those of visible matter. The impact in the mix of the effect of dark energy, however, is more difficult to conceptualize, but crucial.

The most likely candidate for dark energy is energy that's intrinsic to the inherent fabric of space and is often referred to as a cosmological constant. As gravity pulls inward on matter, dark energy essentially pushes outward on space itself.

Famously, when Einstein formulated general relativity, the mathematics naturally incorporated such a cosmological constant whose presence suggested that space *should* expand. At the time though, like Hoyle many years later, he visualized our Universe as unchanging. Uncharacteristically, instead of following the irrefutable logic of the mathematics, he fiddled the books and removed the cosmological constant. Later when it became clear that our Universe is indeed expanding, he called its exclusion the worst mistake of his life.

If in actuality dark energy does act as a cosmological constant, empty space itself embodies energy and of a type whose effect is gravitationally repulsive. As our Universe expands it exerts a force that results in the energy *density* of empty space remaining *constant*—hence the name.

At the big bang, the density of visible energy-matter and dark matter were together enormously greater than the density of dark energy.

As our Universe expands though, their combined energy density has reduced, eventually becoming smaller than that of dark energy. However, the current state of cosmological knowledge points to the fact that overall, while the combined *density* of visible energy-matter and dark energy and matter reduces during the lifetime of our Universe, the total energies remain constant; they are conserved.

The flatness of our Universe and its expansion from the moment of the big bang also embody a further critical and astounding implication.

As cosmologists such as Lawrence Krauss have shown, in a flat and expanding universe such as ours and *only* in such a universe, the attractive and repulsive energies *exactly* cancel each other out to zero throughout the totality of space and at all times.[1] This applies even when the attractive gravitational effects of dark matter and the repulsive energies of the cosmological constant of dark energy are included.

When all its energies and matter are taken into consideration, from the moment of its birth and throughout its lifetime until its demise, the positive and negative forces within space-time continually balance out to zero.

Our Universe is literally, in its totality, formed from *nothing*.

Together, the universal conservation of visible and dark energy-matter and their net zero value enables a dynamic process of evolution to unfold during which each dominates at different epochs in the finite lifetime of our Universe.

From the outward push of the big bang, the expansion of space slowly decelerated owing to the gravitational effect of both visible and dark matter enabling them to cluster into stars and galaxies. But as space extends beyond a certain point, the density of such attractive energies diminishes until a crossover is reached with the density of dark energy then dominating over the remaining lifetime of our Universe.

In 2011 a five-year study, called the WiggleZ project, of some two hundred thousand galaxies by NASA's Galaxy Evolution Explorer (GALEX) ultraviolet orbiting telescope, allied with the Anglo-Australian telescope on Siding Spring Mountain in New South Wales,

Australia, looked out and back over seven billion years. It discovered that this crossover between attractive and repulsive forces actually already occurred some five billion years ago.[2]

Completely unexpectedly, it showed that since then the expansion of our Universe has been accelerating and will likely do so for the remainder of its lifetime. In early 2015 though, astronomers realized that the brightness's of type Ia supernovae, used to calibrate cosmological distances, were different in the early epoch of our Universe. While their consequential recalculation reduces its expansionary pace, importantly it still confirms that the acceleration is real.[3]

In a little while we'll return to further consideration of what the ultimate fate of our Universe may be. But in the meantime, we'll conclude our review of the first law by restating it in terms of the conservation of information.

We've already seen how energy and matter can be restated informationally. Indeed, many information theorists use this equivalence to state that information itself is intrinsically conserved. I believe, though, that their perspective is only partial, owing to their not following through with their understanding of information to its logical conclusion.

Instead, as information is all-pervasive of everything that we call physical reality, where it relates to and is expressed in universally conserved quantities such as energy-matter and others such as electric charge, momentum, and angular momentum (rotational spin), it too is indeed conserved. This is the first law of information (or infodynamics).

But critically, where it is expressed through non-conserved attributes, such as entropy, which we will now explore, then such information too is not conserved.

THE SECOND LAW OF INFORMATION

In the nineteenth century the concept of entropy was first studied as a measure of the order or disorder in thermodynamic systems. This led to its description as the second law of thermodynamics, which states that the entropy of an isolated system will always tend to stay the same or to increase.

As an isolated system our entire Universe embodies this principle of an ever-increasing level of entropy. Its fundamental universal nature gives rise to one of my favorite scientific quotes, by British astrophysicist Sir Arthur Eddington.

> The law that entropy always increases, holds, I think, the supreme position among the laws of Nature. If someone points out to you that your pet theory of the universe is in disagreement with Maxwell's equations—then so much the worse for Maxwell's equations. If it is found to be contradicted by observation—well, these experimentalists do bungle things sometimes. But if your theory is found to be against the second law of thermodynamics I can give you no hope; there is nothing for it but to collapse in deepest humiliation.[4]

Maxwell's equations describe the foundations of electromagnetism, perhaps the most primary of physical forces. That Eddington considered them to be secondary to the immutable rule of entropy and its inevitable increase is, as we'll see, revealed to be crucial to the evolution of our Universe and the flow of time itself.

There are many everyday ways to visualize or experience the increasing flow of entropy.

One, which I'm personally very good at, is knocking something over, usually a cup of tea. Despite my rushed attempts to clear up the disorder, it's clear that I can never return to the pre-spilled order of the cuppa. If someone videoed my clumsiness and then ran the film backward, it would also be clear to anyone viewing what the correct direction of time is throughout the process.

A good way of understanding why this should be so is to picture yourself holding an unused pack of playing cards. Fresh from their packaging, the usual deck is arranged in numerical order from ace to king and so on and by each of the four suits of spades, diamonds, clubs, and hearts. Then imagine yourself throwing the cards into the air (it's more fun though if you actually do this). Whether imagined or real, you'll see that the cards fall in a way that almost inevitably disrupts their initial order to a greater or lesser degree.

If you collect the cards, re-sort them back into their unused state, and then throw them up in the air again, you'll see that they land in a similar but different state of disorder. You can repeat this process as often as you like, but however many times you do so you're exceedingly unlikely to find that when the cards land, they remain in the same sequence as their pre-thrown, ordered state.

This is because while the unused sequence of the pack has just one way of being ordered, there are numerous ways in which the post-throw cards can be sequenced in their disorder. So the entropy of the pack of cards, which as we've seen can simplistically be viewed as a measure of its order and disorder (the lower the entropy, the higher the order, and vice versa), either remains at its lowest level in its initial pre-thrown state or otherwise is almost certainly likely to increase its level of disorder over time.

This apparently simple principle actually offers us a profound insight into the nature of time itself. For, if we rewind time 13.8 billion years to the moment of the big bang, the second law states that the entropy of our Universe at its birth was the lowest it has ever been and will be throughout its lifetime. We can then see how the inevitable increase of entropy from that first moment literally gives time its so-called arrow.

Before we consider other aspects of the second law, let's first debunk a well-known challenge to its universal applicability while at the same time redefining it in terms of information.

In 1867, physicist James Clerk Maxwell (the same Maxwell later referred to in Sir Arthur Eddington's quote about the immutability of the second law) came up with a thought experiment. He envisaged a box of gas, divided in half by a barrier, and an imaginary being, who came to be known as Maxwell's demon, able to control the barrier thereby allowing faster moving gas molecules through while preventing slower moving ones from doing so. Eventually all the faster molecules end up on one side of the barrier, which becomes hotter, even though no energy has been added to the system and in apparent violation of the second law; in this experiment order appears to be retrieved from disorder.

It took a long time to refute this perceived breakdown. Scientists though now appreciate that Maxwell's demon has to measure the speed of all the molecules before deciding which to let through, and this measurement requires energy. When this is accounted for, there's no violation of the second law.

In 2010, Shoichi Toyabe and his colleagues at Chuo University in Japan undertook an experimental version of Maxwell's demon on a tiny scale.[5] Creating a minuscule staircase made from an electric-field gradient, they placed a minute bead on the bottom step. The motion of air molecules around it then naturally jostles the bead and sometimes sufficiently enough to push it up a notch. Videoing the process on a continual basis, every time it went up a step, the team altered the electric field so that the bead couldn't fall back again—equivalent to the barrier in Maxwell's original thought experiment and again seemingly violating the second law.

However, in Toyabe's experiment, there's no conventional energy flow into the system. Instead, by using the video camera to determine the bead's position, the team only used its information as the requisite input to balance the books and preserve the second law. And thus when the camera's energy is taken into account, the entire system obeys the second law precisely.

Armed with a growing appreciation that information underpins the second law too, let's see what further deep insights the law has to offer us. We've already seen the fundamental relationship between energy, information, and entropy and how this universal concept of entropy can be viewed as representing the information content of a system.

So the first moment of the big bang, when the entropy of our Universe was at its lowest, also represented the minimum amount of information embedded within space-time. Since then, while the attributes of information as embodied within the energy-matter balance of our Universe have altered their forms, the total information so manifested remains conserved throughout its lifetime: the first law of information (or infodynamics).

Information associated with non-conserved attributes and primarily entropy, however, continues to inexorably rise: the second law of information (or infodynamics).

It's the second law of information that enables ever-greater levels of complexity to evolve and the development of higher phases of intelligence, consciousness, and self-awareness to be embodied and expressed in our perfect Universe. Indeed, it's this expanded restatement of the second law that enables us to consider space and time themselves and their integration as space-time as being emergent informationally entropic phenomena.

To gain an understanding of how such in-formation may literally be encoded, we now need to return to the minute realm of the Planck scale and the first moment of the big bang.

PIXELATION OF SPACE-TIME

One of the powers of the Schrödinger wave equation that we encountered in chapter 1 is that it predicts that if certain properties of a quantum system are measured, the result may be quantized. Yet, not every measurement gives rise to a quantized state: position, momentum, and time aren't quantized but can take continuous values. A profound implication of this is that space-time may not be quantized, a view that offers clues as to why despite nearly a century of trying, resolving quantum theory and the relativity theory of gravity remains intransigent.

Considering space-time at the level of the Planck scale and restating it in informational terms, though, offers us a major step forward in the resolution of this conundrum.

We first noted the Planck scale as the almost unimaginably tiny or extreme levels at which all the physical forces of our Universe come together. For space-time, this scale is of a spatial length of approximately 10^{-35} of a meter and a temporal length of around 10^{-44} of a second.

Some of you, like me, may be old enough to remember the early days of television, when the resolution of the TV screen was pretty low, and if you went close enough you could see the individual pixels making up the image. Pixels, short for *picture elements,* are the single points or smallest programmable components in a graphic image. Nowadays the development of high-definition media has increased the number of pixels enormously, and the image appears as a continuous visual no matter how close you approach.

Cosmologists are beginning to view space-time itself, rather than being quantized, as instead being pixelated at the Planck scale. Their investigations are also related to discoveries about black holes and, most importantly, their information content. While we'll explore a few other topics first to give us a more comprehensive foundation, we'll return to this emerging understanding in more detail later, as it forms the basis of the most radical and revolutionary aspect yet of the unfolding twenty-first-century scientific perception of our Universe.

For now though, we need to take on board one key finding.

In studying the maximum amount of information and entropy able to be contained within a region of space—essentially the information in bits required to completely describe a physical system at the quantum level—Israeli physicist Jacob Bekenstein derived the so-called Bekenstein's bound.[6]

Some very sophisticated mathematics combining general relativity theory and the second law with the physics of black holes revealed that Bekenstein's bound equates exactly to the entropy of a black hole. In other words, black holes embody the maximum information able to be contained within the region of space that they occupy.

But this isn't Bekenstein's most astonishing revelation. He found that this maximum amount of information for a spherical black hole isn't proportional to the three-dimensional volume of space it occupies but is instead proportional to its two-dimensional surface area. If you read this quickly, may I suggest you reread it and hopefully realize how truly amazing this discovery is.

I've a large textbook that was the bane of my first year undergraduate life at Oxford: B. I. Bleaney and B. Bleaney's *Electricity and Magnetism* (second edition). It measures about seventeen by twenty-four centimeters with a thickness of nearly four centimeters and is stuffed with information, much of it initially virtually indecipherable to me.

If I was to add up the total information contained within it, I'd calculate the approximate average number of bits of data on each page then multiply by the number of pages (equivalent to its thickness). Essentially, given the same density of information per two-dimensional page, when multiplied by the thickness, the overall amount is thus proportional to its three-dimensional volume.

But this is not what Bekenstein discovered. Instead, his calculation of the information embodied by a black hole, or indeed the *maximum* contained by any region of space, is proportional to its two-dimensional boundary. In the case of my book, Bekenstein's bound is proportional to the surface area of its cover, and not its inner volume.

The most incredible implication of this finding, however, goes way beyond books and black holes. For from Gerard 't Hooft's first proposal of the holographic principle in 1993, more and more cosmologists are considering that all the information apparently embodied by the space-time of our Universe may be considered as being imprinted on its boundary.

Bekenstein's bound and much more that's now being studied at the leading edge of science and across multiple scientific disciplines—as we'll return to in much more detail—is coming to the extraordinary perspective that our Universe really is a cosmic hologram.

In beginning to digest this possibility, for which I'll provide much more evidence as we continue, let's first see how it applies to the pixelation of space-time at the Planck scale.

You may have surmised already that Bekenstein's bound measures the maximum number of bits as a multiple of the number of Planck-scale spatial areas making up a larger two-dimensional surface area of a three-dimensional object, where each Planck scale area encodes one bit of information.

Bekenstein's Bound also provides us with an insight into why space itself has expanded since its inception at the big bang. By considering the second law's inevitably increasing flow of entropy throughout our Universe's lifetime, recognizing that at its most fundamental such entropy is informational and also appreciating that it is encoded on its boundary at the Planck scale pixelation of space-time, we can draw a hugely important conclusion: for ever more information in the form of evolution and increasing diversity of experience to be expressed within our Universe, its boundary, space itself, *must* expand—which it has and continues to do.

COMBINING THE FIRST AND SECOND LAWS

In writing this book I came to the realization that restating the laws of thermodynamics to laws of information offers a way of reconciling quantum and relativity theories and provides the instructions that guide the entire course of our perfect Universe from its birth and throughout its life. Combining their universal principles also enables us to gain insights about its likely future and ultimate demise.

To understand how, we need to consider another fundamental attribute that applies both to energy and entropy: the notion of temperature, with one measure of entropy being energy divided by temperature.

So a change in entropy requires either a change in energy, a change in temperature, or both. In our Universe, energy is conserved. So its increasing informational entropy requires the temperature of space-time to change. In thermodynamics, the temperature of a gas will reduce if the space it occupies increases. The entire space of our Universe is expanding (and as we've seen, its doing so enables ever more Planck-scale pixelation of information to be encoded as bits on its holographic bound). So, inevitably, its temperature is reducing.

All this fits beautifully with cosmological observations. Yet there's one circumstance that, at first, seems anomalous. To understand this, we need to go back, once again, to the Planck scale and the first moment of the big bang.

When we include another universal factor, the Boltzmann constant that we first encountered in the calculation of thermodynamic and informational entropy, into the balancing of physical forces, we can derive a Planck-scale temperature. It's a colossal 10^{32} degrees Kelvin (or K for short). To get some idea of the size of this number, the Kelvin scale begins at absolute zero, ice melts under normal atmospheric conditions at 273 degrees K, and the central temperature of our Sun is a mere 16 million degrees K. The enormous Planck temperature is that of the very first moment of our Universe at the big bang.

And the wavelength of its radiation is one Planck length.

With the pixelation of space-time embodying a single bit of information per Planck-scale area, the Planck-scale genesis of our Universe encoded its primordial information expressed at Planck-scale wavelengths of light, essentially embodying the simplest instructions possible to switch on and light it up. The apparent anomaly, though, is that the energy associated with such a massive temperature should be expected to be violently chaotic, and so instead of its being at its lowest level of ordered entropy, our early Universe should have incorporated its greatest level of entropic disorder.

Without the presence of gravity or indeed the extremely strong electromagnetic field of this earliest epoch of our Universe, such a high temperature, usually associated with equivalently high kinetic energy, would indeed have been a frenzied firestorm of energies.

But as we're discovering in some of the most extreme astrophysical phenomena we're able to observe, enormously high levels of gravity and powerful magnetic fields are able to tame and order energy-matter at such extreme temperatures. As such, the energies at this initial Planck era are ordered with only very small variations that would later form the gravitationally attractive seed-points for stars, galaxies, and other large-scale structures throughout our Universe.

After 13.8 billion years, the average temperature of our Universe has now reduced to a very chilly 2.7 degrees K based on measurements of the cosmic microwave background. As space continues to expand, its average temperature will also continue to fall.

The logical conclusion given our Universe's entropic progression, flatness, isolation, and finiteness suggests that its end will be as a state of maximum informational entropy and complete thermal equilibrium at or very close to absolute zero. By combining cosmological measurements of the accelerating expansion of space-time, the rate of cooling, and a calculation for the maximum informational entropy possible at absolute zero, scientists are coming close to being able to estimate the likely timescale before this occurs. Even when they do though, there's still no known physical mechanism for the final dissolution of our perfect Universe, for how it will end its life.

However, perhaps like a bubble that bursts as its inner pressure equals that of the surrounding atmosphere, this end-time may express a point of equivalence when our Universe may release its accumulated information, knowledge, and wisdom into the infinite cosmic plenum within which it was born, lives, and will die.

Indeed, as we perceive the extraordinary lifetime of our Universe from this new perspective, we may indeed come to see it not as beginning in a big bang but born and evolving as a "big breath." From now on we will describe it from this viewpoint.

3
Conditions

Initial circumstances that determine the outcome of something . . .

One of the first conditions of happiness is that the link between Man and Nature shall not be broken.

Leo Tolstoy, author

Simplicity, invariance, and causality are three fundamental conditions of our perfect Universe. Each plays a major role in how space and time itself is manifested and how embedded information in-forms physical reality.

The realization of the astonishing level of simplicity that underlies the complexity of the physical world often shows up in attempts to integrate apparently diverse phenomena. Sometimes simplicity is found in unexpected revelations of correspondences and complementarities. Regularly its presence guides the path of least effort or resistance that shows up universally in physical processes.

The universal condition of invariance, meaning unchanging or constant behavior, states, for example, that while space and time, as Einstein discovered, are each relative to the position of an observer, their integration as four-dimensional space-time is invariant. So, all measurements of the same event as defined by its coordinates in space-time will give the same answer to every observer, regardless of the observer's location in space or time.

Indeed, the mathematics of relativity theory actually requires that throughout space-time, *all* physical laws must be invariant. They must take the same form regardless of whether they're measured in a laboratory on Earth today, on a planet orbiting a star in another galaxy a million years ago, or even at the edge of a black hole a billion years hence.

The third primary condition is causality. This innate prerequisite of the physical realm always ensures that any cause precedes its effect. This may just seem like common sense as it inevitably reflects our everyday experiences, yet many physical laws appear not to include time and thus the notion of cause and effect. As we look deeper, though, we'll find that, vitally, causality is built into the fundamental structure of space-time.

These three crucial conditions determine and pervade physical reality. Let's now see how they do so and how they are key aspects in enabling our Universe to exist as a unified and coherent entity and evolve to embody ever-greater levels of complexity.

SIMPLE

Let's first appreciate the view of Galileo Galilei, the sixteenth-century astronomer and early champion of the Copernican view of the Sun being at the center of our Solar System, that the book of nature is written in the language of mathematics.

The basic mathematical functions, which underlie and are encoded in the quantized bits of physical reality, haven't been invented, constructed, or created by human intelligence: they've been discovered. And as a literally cosmic language, all the physicality of our Universe and so its all-pervasive information can be conveyed in mathematical terms that describe its relationships, transferences, and flows.

While all physical reality can be expressed informationally in the cosmic language of mathematics, not all mathematics describes such manifestation. Time and again scientists have realized that the powerful and elegant equations and relationships that describe physical reality are indeed, to paraphrase Einstein's view, as simple as they can be but no simpler.

Being guided by the condition of simplicity and willing to follow where its descriptive mathematics leads reveals further insights that enrich and integrate our understanding of the nature of our perfect Universe. Its direction enables us to distinguish the apparent diversity of phenomena from the deep simplicity that underpins their expression. To get a feel for how important this premise of innate and universal simplicity is, let's briefly consider a few examples.

In the late nineteenth century Scottish physicist James Clerk Maxwell, often called the father of modern physics and an all-around good egg, studied the seemingly diverse natures of electricity, magnetism, and light. He built on research by one of the best experimental physicists of his or indeed any era, Michael Faraday, who in the 1830s had begun the process of simplification by discovering that a moving electric charge causes a magnetic field and correspondingly that a moving magnetic field generates an electric current.

This led Maxwell to realize that rather than being separate, electricity and magnetism effectively mirror each rather like two sides of the same coin. By depicting them in the simplest of mathematical terms, he was able to unify them as electromagnetic, or EM, fields. The mathematics also revealed that EM fields travel in waves through space at the speed of light, which he thus realized is a form of EM radiation whose wavelengths are visible to us.

His powerful depiction of EM fields was able to predict the existence of radio and other non-visible wavelengths of the EM spectrum. Not only though are his equations the basis for our global computing, communications, and many other technologies, but they also formed the inspiration for Einstein's later revolutionary insights into the fundamental nature of space and time.

In the early twentieth century, by taking literally Maxwell's mathematical description of light that showed it as having a universally fixed speed, Einstein realized that both space and time are relative to the perspective of an observer. By integrating them into a four-dimensional space-time, he was able to describe physical reality in an invariant way and one that always preserves causality.

Where the mathematics that accurately describes a phenomenon seems, however, to be less simple than it could be, a door is invariably opened to a more profound and compelling understanding, revealing why it's as simple as it is *but no simpler*. Indeed, throughout the history of science, such recognitions have always shown that the actual level of simplicity is minimized to the degree that's required to still enable the existence and evolution of our perfect Universe. By testing the limits of simplicity, science has shown that, at its most fundamental, everything in our Universe is purposeful, and nothing is wasted.

An illustration of this is English theoretical physicist Paul Dirac's equation describing the behavior of elementary particles, which in the 1920s led him to predict the existence of antimatter, which was subsequently confirmed four years later. Antimatter has the same mass and is also made of ordinary matter, but it has opposite electric charge.

Dirac's equations showed that there should have been the same amount of antimatter and matter created at the beginning of our Universe, yet antimatter is currently barely present. When a matter and antimatter particle come together, they annihilate each other in a burst of light. Measurement of light photons and matter within the CMB has shown that by about a billionth of a second after the first moment of the big breath, for every billion or so matter-antimatter pairs that destroyed each other, only a single matter particle survived.

The ongoing exploration to understand why the initial "complication" of antimatter was needed, why the initial simple symmetry of matter-antimatter was broken, and why only one in a billion matter particles was left is helping us answer a fundamental question of why there is something, other than light, in our Universe. Once again this quest is being guided by the principle of being as simple as it could have been but no simpler.

Another integration aiming to simplify a phenomenon and so understand it at a deeper level was achieved by Argentinian physicist

Juan Maldacena. At a 1998 conference of string theorists, he presented a paper inspired by the concept of the holographic principle, which as we've seen was pioneered by 't Hooft and Bekenstein, among others.[1]

He disclosed that quantum fields can be restated as a version of M-theory (which itself seeks to reconcile quantum and relativity theories) defined on a holographic boundary. In revealing their correlation, he enabled the holographic principle as a profound and revolutionary new perspective to gain much greater acceptance among physicists. As an added benefit, he showed too how a very difficult problem viewed from one perspective is much simpler when considered from its correlate, although his great insight is usually known by the decidedly unsimple title of the AdS/CFT (anti-de Sitter/conformal field theory) correspondence.

We'll soon explore in more detail the key breakthroughs, enormous implications, and rapidly expanding perspectives of the cosmic hologram, but for now suffice it to say that Maldacena's presentation was so well received by his colleagues that they immediately began a celebratory dance based on the macarena.

SPACE AND TIME

We can better understand how the two other fundamental conditions, invariance and causality, shape the existence and evolution of our Universe on large scales by now exploring further the nature of space and time, although here too simplicity will be our guide.

As we do, we'll see why space and time need to be relative to the perspective of an observer while naturally integrating into invariant space-time, and how causality is always preserved owing to the finite and constant speed of light.

Albert Einstein was sixteen years old when he first followed a beam of light in an imaginary thought experiment. It took him a further decade though, until 1905, his so-called year of miracles, to publish a series of breakthrough discoveries inspired by what he'd first intuited and which culminated in his special theory of relativity.

The "special" theory is so called because it confines itself to describing the special case of the motion of bodies moving at a *constant* velocity in relation to each other. As we'll see soon, it took even Einstein's genius a further ten years of intense thought to extend his understanding to the so-called general theory of relativity that describes the motions of accelerating bodies. Oh, and along the way, to reveal a revolutionary new theory of gravity.

But to get a grip on all this, let's begin by returning to the start of Einstein's journey of discovery to examine what I consider one of, if not the, greatest insights yet achieved in the history of science, one that's completely contrary to our everyday experience.

To explain how radical his conclusions are, let's remember the basic lessons we learned at school about the relationship between two bodies traveling at constant velocities. Specifically, the velocity of a body is its speed in a given direction. So to keep things simple, we'll just consider bodies moving in parallel or opposite to each other. The rules of motion then state that to arrive at their combined velocity, their velocities should simply be subtracted from each other if they're moving in the same direction, or added if they're moving in the opposite direction.

We can clearly appreciate and experience this if, for example, we're in a car, say traveling in the middle lane of a three-lane highway at a constant velocity of sixty-five kilometers (about forty miles) per hour. On the inside lane is a slower truck traveling at, say, sixty, and in the outside lane a sports car traveling at seventy.

To determine our relative velocities is easy. The relative velocity between us and the slower and faster vehicles is five kilometers per hour in each case. And the relative velocity between the sports car and the truck is ten kilometers per hour.

On the other side of the central barrier, say a vehicle is traveling in the opposite direction to us at the same average velocity of sixty-five kilometers per hour. While safely separated by the barrier, our relative velocity is a whopping 130 kilometers per hour, as in this case our velocities are additive.

All this is so in tune with our common sense that as drivers we don't tend to give it much conscious thought as we weave in and out of traffic. But as Einstein discovered, light doesn't operate in this commonsensical way. And realizing this when he scrupulously followed where the logic of Maxwell's electromagnetic equations led, he was then forced to acknowledge that neither does space nor time.

To take our next step in understanding what he discovered and its implications, we first need to define what we mean in our everyday example by our sixty-five-kilometers-per-hour velocity as we travel along the highway.

The speedometer that tells us that this is how fast we're traveling measures the car's motion in relation to the road and thus the Earth. Our usual and fortunate experience is that when we stand still the Earth isn't moving in relation to ourselves: otherwise we'd get pretty dizzy. Vitally, it thus forms a stationary reference point not only for our car and every other vehicle on the highway but also for all objects on or closely above its surface.

Now let's turn our attention to Einstein's thought experiment, when he imagined what it would be like to travel toward, away from, and alongside a light beam. As we've seen, some years before him, Maxwell had realized that light is an electromagnetic wave. Critically, his equations that describe EM fields show that within a given medium, whether space or water, for example, the speed of light is a constant, always. So although light travels fastest in the vacuum of space and slower in all other media, within each its speed is invariant.

But the equations do not offer any clues as to what the constancy of light speed is measured *against.* This again is strange when considered in terms of our everyday experiences where speed is always measured against some frame of reference.

In our earlier example, the velocity of the vehicles on the highway is measured against the fixed stationary reference frame of the Earth. But this isn't the case for light that travels throughout our Universe and so, if it behaved in the same way, would need a universal standard frame of reference for its measurement.

For many years an all-pervasive ether had been proposed, not only to try to account for such a nonmoving reference frame but also to answer a related mystery: What medium does light travel through in the apparent vacuum of space? Despite numerous efforts to detect it, by Einstein's day (and ever since) scientists had been continually unsuccessful in doing so.

Let's first take a look at the easier of the two enigmas: What medium does light travel through in space?

Until Einstein and with no clue from Maxwell, physicists had viewed the propagation of light as needing some form of medium to travel through. As mechanical waves, sound needs some form of medium, such as air or water to spread; as the famous tag line from the film *Alien* noted: "In space no one can hear you scream."

The transmission of visible light (and all other wavelengths) of EM fields doesn't require any such media, because, as Einstein realized, light is an oscillation of the EM field, and the EM field itself *is* the medium, universally present throughout space-time.

The greater challenge and correspondingly even more revelatory discovery arose from Einstein's attempt to answer the other puzzle: What is light's universal reference frame? When Einstein's thought experiment followed a beam of light, he realized astonishingly that the constancy of its speed means that it's constant for *any* and *all* frames of reference no matter how they're chosen. Essentially, from wherever and however it's measured, its speed will *always* be the same.

When considered against our everyday experience, this is extraordinary.

To see just how unexpected this discovery was, we'll consider another everyday example. Imagine we're traveling in a train on a dark night when another passes by. Unable to see the surrounding landscape for reference, as the two trains pass, we'd be unable to determine which one, or if indeed either, was stationary. We wouldn't be able to gauge the absolute speed of either against a stationary reference point. All we would be able to measure would be their *relative* speed.

If each train had a mirror positioned on its leading edge, we could do this by sending a series of light signals from our train and measuring the progressively shortening times it takes them to reach, bounce off the other train's front mirror, and return.

Say the oncoming train is traveling at sixty-five kilometers per hour and our train (without our knowing) is stationary, we'd measure the oncoming train's speed at sixty-five. Now if our train was traveling at forty kilometers per hour (again without our knowing), we'd calculate that the other train is approaching at 105 kilometers per hour. And yet in each case, the oncoming train's speed remains the same.

And if our trains were running on parallel rails in the same direction (and with the other train also having a mirror located on its back end), while being able to measure our relative speed of twenty-five kilometers per hour, again we wouldn't be able to tell either of our absolute speeds.

However, if instead of being a train, imagine a beam of light is traveling toward us, though of course much faster. *Regardless* of our own speed and unlike an approaching train or indeed any other material object, we'd *always* measure the speed of the light as a constant amount. Equally, if a beam of light is traveling away from us, our measurement of its speed would be exactly the same, irrespective of how fast we ourselves are moving.

While traveling through a medium such as air, light travels more slowly than in a vacuum; nonetheless it will always be measured as propagating at a constant rate. In the vacuum of space, we'll always measure the speed of light at its universal (and maximum) speed limit of nearly 300,000 kilometers per second, denoted by the letter *c* (originally short for *constant* and more recently for *celeritas,* the Latin for "swiftness").

Einstein's conclusion that throughout the Universe there's *no* preferred location or direction throughout space and that all *inertial* frames of reference (meaning those traveling at a constant velocity relative to each other) are completely equivalent also reveals that the laws of physics are the same throughout our Universe.

This is good news. Without the implications of the constancy of light as described by the relativity of space and time, we couldn't develop any cosmological theories that would apply to our Universe as a whole. Light and its constancy, as Einstein went on to discover at even more profound levels of reality, enables our Universe to exist and evolve as a coherent and unified entity.

This universality of physical laws, though, has a deeper and unavoidable implication. It took Einstein's genius to appreciate that the constancy of light speed as measured by all observers means that the speed they are traveling affects their observations and measurements of space and time themselves.

The velocity of an object is measured as the distance it travels divided by the time it takes to do so; for example, sixty-five kilometers per hour. As an object speeds up, Einstein's special relativity theory revealed that time (for example, as measured by a clock) slows down, or dilates, and distance (for example, a measured length) along the direction of movement shrinks in perfect correspondence. So, any measurement of the speed of light will remain exactly the same and identical to that made by every other observer regardless of their different comparative speeds.

These precisely compensating relationships are mathematically described by the elegantly simple Lorentz transformation equation, named after Dutch physicist Hendrik Lorentz, which also applies to the mass-energy of an object. It shows that if a massive object were able to be accelerated to the speed of light, time would stand still, its length would shrink to zero, and its mass would increase to an infinite level. The energy to increase its velocity to light speed would also be impossibly infinite, which is why no object of any mass can ever do so within space-time, and why light itself can only travel at the universal speed limit of *c,* because it has no mass.

SPACE-TIME

Owing to the relativity of space and time, different observers will measure the same event from different perspectives. Their relativity, however, preserves the constancy of light speed and enables physical laws to be universally applicable. This finite and invariant speed of light is a universal conversion factor that changes units of time into units of space and vice versa. It literally weaves space and time together into a four-dimensional and invariant entity called space-time—a further revelation that our Universe is innately interconnected and exists and evolves as a single unified entity.

To understand the concept of invariant space-time better, let's first look at how distances in space-time are measured. Although most of us don't usually consider such space-time distances on an everyday basis, we're familiar with one term that does just this: a light-year, which is the distance that light in a vacuum travels in a year.

Combining the speed of light multiplied by time gives us a distance. So our nearest star, the Sun, is eight light-minutes away, whereas Andromeda, the nearest major galaxy beyond our own Milky Way, is more than 2.5 million light-years distant.

Because nothing can exceed the speed of light within space-time, when we see sunlight, we're actually seeing the Sun as it was eight minutes ago, and when we use a telescope to view the light from the beautiful Andromeda Galaxy, we're seeing it as it was more than 2.5 million years in the past.

CAUSE AND EFFECT

Owing to the very special flatness of space in our Universe, we can use the principles of geometry that we first learned at school not only to derive distances in space-time but also to discover a mathematical basis for how the nature of invariant space-time ensures that all events within our Universe always maintain causality.

With the relativity of space and time, different observers can each correctly argue from their own perspective about whether or not two events in different locations take place simultaneously. However, when considered as events in combined space-time, any disagreement is resolved as every observer will measure the same result.

An alternative approach to understanding how causality is sacrosanct in our Universe comes from taking a deeper look at the nature of time itself, which we'll now do.

WHAT *IS* TIME?

We've previously noted that the entropy of our Universe began in its lowest state—in informational terms, the simplest it could be. As entropy has inevitably increased since, this essentially gives the notion of time its so-called arrow, or direction.

We've also begun to explore how cosmologists are beginning to view space-time as being pixelated at the minute Planck scale, and how space, rather than our apparent experience of three dimensions, can instead be informationally depicted as integrated space-time as a two-dimensional holographic boundary and where each Planck-scale area embodies one bit of information. We saw too that the combination of the inevitable increase of informational entropy and such a holographic approach to space-time requires space itself to expand.

The corresponding conclusion with regard to time brings us via a different route from that of Einstein's journey to special relativity to the innate preservation of causality. In informational terms a causative event always embodies less information than the effect arising from it, as the effect has to take account of its cause as well as its consequence. In addition, the inevitable flow of entropy means that for each succeeding Planck timescale of 10^{-44} seconds our Universe expresses ever more information as the past in-forms the present, which then in-forms the future. So at all times causality plays out within space-time, from the beginning to the end of our Universe's lifetime.

Indeed, we may now consider the nature of time as experienced within the physical realm, *literally* as the informational flow of entropy.

NONLOCALITY

Despite ever more sophisticated attempts to do so, most scientists have pretty much given up on trying to exceed the speed of light within space-time and thus disprove Einstein's revelation that it represents a universal speed limit for transmitting any information within our Universe.

However, in the past few years quantum physicists have used an aspect of the phenomenon of nonlocality to develop quantum computers potentially capable of processing gigantic amounts of data dramatically faster than our current technology allows.

Nonlocality, which Einstein famously derided as being "spooky action at a distance," was an apparent consequence of quantum theory that he never accepted. In 1964, Northern Irish physicist John Stewart Bell, however, in a rigorous mathematical proof named after him, showed that nonlocality is required to reproduce all the physical attributes of quantum theory. Experiments beginning in the 1970s, aimed to validate nonlocality, and in 1982, nearly thirty years after Einstein's death, Alain Aspect and his colleagues at the Université d'Orsay in Paris proved conclusively its reality.[2]

Nonlocal connectivity is often demonstrated through the creation of pairs of twin particles that share quantum states and behave as a single entity—an effect known as entanglement. If they're then separated and the state of one of the entangled pair is subsequently altered, the other will *immediately* switch its own state to mirror that of its twin, regardless of how far apart in space and time they are.

Observing and so gaining information about either or both particles of the pair disturbs them and upsets or destroys their entangled state. In 2014, in a so-called quantum teleportation experiment, a team of physicists at the University of Geneva achieved a remarkable feat that further demonstrated the informational significance of entanglement.

They did so by first creating an entangled pair of photons; the first was propelled along an optical-fiber cable some twenty-five kilometers (about 15.5 miles) in length and the second sent to a crystal close to where the pair was created. A third, unentangled photon, was then targeted to hit the first, obliterating both of them.

Here's when the quantum teleportation occurred. The *information* contained in the third photon, instead of being destroyed, was conveyed in the crystal that also contained the surviving second photon of the previously entangled pair.

Félix Bussières, the lead author of the report describing the experiment, explained that "the quantum state of the two elements of light, these two entangled photons which are like two Siamese twins, is a channel that empowers the teleportation from light into matter."[3] The experiment thus proved that the *information* of the state of the photons takes precedence and is thus more fundamental than its physical expression.

Progressively, experiments have proved that the phenomenon of entanglement extends way beyond the quantum scale. Beginning with entangled pairs and even triplets of photons, extending to entangled electrons and large molecules, in 2011 a team from the University of Oxford led by Ka Chung Lee and Michael Sprague even managed to entangle the quantum states of small diamonds each the size of an earring stud and at room temperature.[4] At billions of times larger than the quantum scale and at a temperature we experience in our everyday lives, such correlation shows that it's a reality in the macro-world and at the scales at which relativity theory also applies.

What the experiments have shown is that both quantum and relativity theories are correct in that while such nonlocal connections are real and transcend the limitations of space and time, no information is actually transferred *within* space-time, and light's universal speed limit remains inviolate.

As highly technical investigations have shown though, entanglement is merely a precedent to universal nonlocality, which, as Bell's theorem essentially shows, is fundamental for *all* the phenomena predicted by quantum theory and that have now been overwhelmingly validated by experiments.

INTEGRATION

Combined with the inviolate nature of space-time, universal nonlocality offers us profound evidence that our Universe is fundamentally interconnected as a unified entity that is underpinned and permeated by information. The universal speed limit exhibited by light ensures that information is transferred at a constant and finite limit within space-time, maintaining causality and enabling our Universe to experience and evolve.

Yet the innate presence of nonlocality enables the holographic boundary that is essentially the in-formational foundation from which the appearance of the physical world arises to be wholly and simultaneously integrated.

This emerging view also offers an alternative approach to understanding the very special and yet crucial nature of the flatness of space-time and the incredible level of universal homogeneity of our Universe. Instead of being "problems" that have generally been assumed to be resolved by an early inflationary epoch—despite no convincing mechanism for how such a mechanism could occur, how it would end, and its apparently unstoppable consequences in terms of infinite other universes forever branching off from it, and, critically, no evidence for any such occurrence—the inflationary premise has become a mainstay of standard cosmology.

The emerging perspective of the cosmic hologram, however, offers another approach to the reality of universal flatness and homogeneity that precludes the requirement for any inflationary process. In considering all that we call the physical reality of our Universe as arising from an integrated holographic boundary, flatness is encoded as an in-formational basis for the appearance of space-time, and universal nonlocality ensures the requisite level of homogeneity from which evolutionary processes can emerge. Such exquisitely tuned information has literally in-formed our perfect Universe from the beginning and will do so until its end.

But for now, we've seen that inviolate space-time integrated with universal nonlocality engenders the innate simplicity, invariance, and causality that are the fundamental conditions for our perfect Universe to experience ever greater levels of complexity and the emergence of self-awareness.

4
Ingredients

Things that are combined to make something greater . . .

> *The ambition of every good cook must be to make something very good with the fewest possible ingredients.*
>
> Urbain Dubois, chef and writer

There is only a single ingredient that makes up our perfect Universe: information, expressed as energy and acting through non-entropic and entropic processes.

As we've seen, informational entropy increases through the lifetime of our Universe. Information expressed as energy though is not only universally conserved while continually changing its forms, but the flatness of space entails that within the entirety of space-time the combined total of positive or repulsive energies and negative or attractive energies is always precisely zero.

We'll now explore the truly wonderful interactions through which the various energies of our Universe combine in myriad ways to in-form and evolve the manifest world. Along the way we'll discover more about the equivalence of energy and matter; the notion of mass and gravity, and how they interact with space-time; and more about mysterious dark matter and dark energy.

Significantly, we'll discuss how various emerging theories attempting to reconcile quantum and relativity theories all seem to require the reduction in the number of spatial dimensions from the three we're familiar with to a single dimension that combines with time to form a fundamentally two-dimensional space-time—another indication of the holographic foundations of physical reality.

WHAT IS ENERGY?

While there are many forms energy can take in our perfect Universe, all types are convertible into all others, all obey universal conservation throughout space-time, all ultimately balance each other out to zero, and all energy flows adhere to universal paths of least effort.

For example, the EM radiant energy of the Sun, mainly in the frequency range of the visible spectrum, is converted on Earth by plants and stored as chemical energy in compounds such as sugars and starches. When we then eat plant tissue our digestive process breaks down the plant compounds and converts the released chemical energy into other forms such as electrical energy that powers our nervous system and mechanical energy that enables us to move around.

Sometimes heat and work (the application of force to move something, such as swinging a bat and hitting a ball) are mistakenly considered to be energies. Instead, these are actually processes by which the energies of a system are transferred or changed in some way and where the flow of time matters and so actually represent flows of entropy.

Although this distinction might seem a bit arcane, it's actually very important for our understanding of not only what energy is but also how it complements the concept of entropy. The attributes of both, as we've seen, can be understood and restated in terms of the first and second laws of information that are the fundamental instructions informing the physical world.

Where information is expressed by the myriad forms that energy can take, it is universally conserved; whereas information expressed through entropic processes is always increasing throughout the lifetime of our Universe.

EQUIVALENCE OF ENERGY AND MATTER

The most famous equation in the world is reckoned to be the one discovered by Einstein in which he revealed the equivalence of energy and matter: $E = mc^2$, where E is the energy associated with a mass m, and c^2 is the speed of light squared.

Given that the size of c^2 is enormous, the equation shows the incredible amount of energy stored in matter. For example, the energy embodied in half a kilogram (about one pound) of matter, if totally released, could provide all the power needs of a midsize city such as Nottingham in the UK or Austin, Texas, for a year.

While Einstein arrived at their correspondence from his realization of the relativity of space and time and their integration into an inviolate space-time, they're also an inevitable consequence of the universal conservation of energy and momentum and the flatness of space. As we've seen, the early quantum physicists understood that as a consequence of their intrinsic waveform nature and equivalence, *all* energy and so *all* matter have a vibrational frequency; the higher their frequency the greater the energy. So X-rays with smaller wavelengths and higher frequencies are more energetic than radio waves with longer wavelengths and lower frequencies. And matter too, given its equivalence with energy, can also be described in terms of its frequency.

The equation that defines this is $E = h\nu$. E represents the energy of something and is equal to h, the so-called Planck constant (a universal factor that represents the quantum of action—how packets of energy act through time at the quantum scale) multiplied by ν, its frequency. So, the energy of *anything,* whether denoted as a form of energy or its derivative matter, can be denoted by this universal equation, which is literally as simple as it could possibly be.

Finally, as matter can always be restated and described in energetic terms, and as energy can be restated more fundamentally in terms of information, so too matter is inevitably informational in nature.

WHY MATTER MATTERS

Without mass, everything within space-time would travel at the cosmic speed limit of *c*, as do of course the massless photons of light. The Lorentz transformation equation describes the slowing down of time as light speed is approached so that, at this universal speed limit, time literally stands still.

The same Lorentz equation also reveals that if accelerated, the mass of an object increases to infinity at the speed of light: an impossibility that ensures massive bodies are unable to achieve light speed within space-time.

The Large Hadron Collider, or LHC, at CERN, the European Center for Nuclear Research, which restarted in spring 2015 after a two-year shutdown and upgrade, routinely has to deal with such relativistic effects. Its immense power accelerates elementary particles to only a few meters per second slower than light speed, both enormously increasing their masses and, for normally short-lived particles, extending their lifetimes by many thousands of times.

Very importantly, obtaining mass slows things down and, together with the equivalence between mass and energy, essentially enables the entropic flow of information and so the experience of time itself within our Universe. For many years, though, the way in which elementary particles gain mass was a mystery, as there's no indication in quantum theory as to why or how they should do so.

In 1964, British physicist Peter Higgs was one of six theorists who attempted to answer this fundamental question. Proposing the existence of a universal field of constant energy pervading all of space, whose interaction with certain elementary particles imbues them with mass, the field and its mechanism to instill mass is now named after Higgs.

At the extremely high temperatures of our Universe less than a trillionth of a second after the beginning of space-time, all elementary particles are deemed to have been massless. As the temperature dropped though, at a certain point, the falling energy of the Higgs field underwent a so-called phase transition (such as what happens when water freezes into ice). In doing so, the field fell to its lowest energy level, essentially "freezing" that level and the triggering of its associated mechanism for inducing matter with mass into the fabric of space-time from then onward.

In 1993 the then UK science minister offered a prize (a bottle of good champagne) for the best nonscientific explanation of the Higgs mechanism, which was won by David Miller of University College London. Since translated into other similar scenarios involving politicians or Hollywood actors, the original envisages a roomful of scientists having a party. When a well-known scientist enters the room, she causes a bit of a kerfuffle, attracting admirers and finding her progress slowed down as a result. She acquires "mass" due to the "Higgs field" of enthusiasts clustering around her. When a lesser-known person enters, however, fewer people are attracted to him, and so his interaction with the field is smaller and he acquires less mass.

The Higgs field and mechanism is an essential component of our Universe. Not only does it provide the mass needed for time to unfold, but without it atoms would immediately disintegrate. Yet the field's electric-charge neutrality also means that it doesn't directly interact with the universal EM field that pervades all of space. So photons remain massless, which, as we've seen, is important to enabling the attributes of light to literally weave space-time together.

While both the theoretical field and the mechanism by which mass is derived came to be named after Higgs, for many years they remained a merely theoretical possibility, as the Higgs field is very difficult to detect. The only means of doing so, at least for foreseeable technology, is through the creation of excitations within the field, which are predicted to present themselves as an elementary particle called the Higgs boson, and which then acts as a mediator imbuing other elementary particles with mass.

The search for the Higgs boson, however, requires very high energies, equivalent to reenacting those at the earliest epoch of our Universe to excite the Higgs field and enable the incredibly unstable and short-lived particle to appear. Even then, it's only possible to detect it indirectly by the slew of lower mass-energy particles it is predicted to decay into.

The only place on Earth where the required high-energy conditions can currently be attained is at the LHC at CERN. It was here in 2012 that the Higgs boson was first discovered by slamming together high-energy beams of protons and observing the decay-particle outcomes. With a mass of around 130 times that of a proton, the Higgs boson was found to be in line with theoretical predictions, and its discovery led in 2013, nearly fifty years after their breakthrough insights, to Higgs and his colleague François Englert being awarded the Nobel Prize in Physics.

BUILDING BLOCKS

The building blocks of non–dark matter are the elementary particles that make up the so-called standard model of quantum physics. Developed over eighty years, the model is extremely accurate in describing the behavior of quantum systems. Significantly though, its excellent fit is owing to the model's being primarily empirical, assembled from an enormously wide range of experimental findings rather than as the outcome of a theoretical framework. Accordingly, it includes many factors that are specifically input as the results of experimental measurements, rather than naturally emerging from a theoretical context, and it is largely unable to make predictions that could lead to future discoveries and deeper insights. Limited by its lack of theoretical underpinnings and predictive power, it's also unable to account for dark matter or understand why the elementary building blocks of non–dark matter are as they are.

Truly elementary particles are those that are deemed to have no further inner structure. These comprise electrons and neutrinos (together described as leptons owing to their small masses) and quarks, which make up the protons and neutrons of atomic nuclei. Thanks to ever-more powerful particle smashers that, as at the LHC at CERN, train beams of electrons or protons at very high energies together and analyze the crash results, a veritable glut of some two hundred particles have been discovered. The great majority of these though consist of composite or extremely short-lived entities and really represent the higher energetic states or combinations of the more fundamental quarks and leptons.

The standard model recognizes twelve such truly fundamental particles of matter, which are themselves subdivided into a number of three-family groups with similar properties, with each formed of two types of quarks, electrons and neutrinos.

Given that one of our guiding precepts through the exploration of our perfect Universe has been the adherence to the principle that it's as simple as it can be but no simpler, you may ask, why there are twelve such fundamental matter particles?

The answer, as yet, is that no one really knows. There are, though, a number of clues that emerge from considering their individual attributes. Two of the twelve are the so-called up and down quarks with the same mass but different partial electric charges. The up quark embodies a two-thirds positive electric charge and the down quark a one-third negative charge. In combination, two up and one down quarks form the positively charged proton. Similarly, in combination, one up quark and two down quarks form the neutrally charged neutron (unusually, a name that actually "does what it says on the package" and makes sense to non-physicists). Another is the negatively charged electron, whose negative charge exactly balances the positive charge of the proton.

These three fundamental particles, the up and down quarks and the electron, represent the lowest energy-mass representations of their particular characteristics. As we've seen with the Higgs field and indeed is the case for all physical principles, such lowest-energy manifestations by far are the most stable and form the vast majority of the non-dark energy-matter of our universe.

Using an analogy to music, these three basic particles can be viewed as fundamental "notes" of matter. Each has two levels of higher quantized energy-mass equivalents (the so-called charm and strange, top and bottom quarks, and the muon-electron and tau-electron) that may then be understood as essentially higher octaves of the same basic notes and together make up nine of the twelve fundamental particles.

The final three particles making up the twelve are three types of neutrino. These, however, instead of representing fundamental and higher-octave energetic versions of the same particle, have been discovered to continually and harmonically oscillate among the three forms. Again to use a musical analogy, they can be viewed as three variations of a single note: natural, sharp, and flat.

So we may now appreciate the twelve particles of the standard model as forming one fundamental electron and two fundamental quarks (and their six overtones) and three harmonic neutrinos—as simple as can be but no simpler.

We're not quite finished with the building blocks of energy-matter as we also need to consider the fundamental forces of our Universe and the mediating particles through which they interact.

Physicists recognize four such forces. The first is electromagnetism, or EM, the means by which light universally pervades space-time and whose attributes are crucial for the mechanisms of the cosmic hologram. The second and third are extremely short-range forces that apply on atomic and subatomic scales: the strong force that holds together protons and neutrons in atomic nuclei, and the weak force that governs nuclear fission and radioactivity. In the very early high-temperature epoch of our Universe, EM and weak interactions were combined in what is known as the electroweak force until, as energies fell below a certain threshold, their symmetry was broken and the two forces separated out.

The EM force, as we've seen, is mediated by the photon. The strong and weak forces are mediated, respectively, by particles known as gluons (good name for a very sticky force) and W and Z particles (less well named in my view).

As an aside, while as we've seen interaction with the Higgs field gives the elementary particles such as electrons and quarks their mass, for composite particles such as protons and neutrons the majority of their mass (and thus the bulk of visible matter in our Universe) actually comes from the binding energy of the gluons that hold their composite quarks together.

To now understand the differences between matter and force particles we need to first look at an intrinsic aspect of all fundamental matter and force particles: spin. Spin is the quantum equivalent of angular momentum, the amount of rotation something has, such as Earth spinning on her axis.

Unlike large-scale objects that can transfer angular momentum (but only so long as the entirety of it within a system is conserved), the intrinsic quantum spin of particles is exactly the same for every particle of that type and unchangeable regardless of what happens to it. Each type of particle embodies and is assigned a specific spin quantum number. For the fundamental particles of matter, such spin is always a half integer—1/2, 3/2, and so on—whereas for force particles the spin is always an integer.

It turns out that when the relativity of space-time and quantization of energy-matter are taken into account, there can indeed only be two such types of particles. All the half-integral spin particles that make up matter are collectively known as fermions (after the Italian physicist Enrico Fermi). Those that have integral spin are called bosons (after Indian polymath Satyendra Nath Bose). The basic difference in spins between fermions and bosons means that matter and force particles behave very differently—which is just as well, as otherwise our perfect Universe would not exist.

We've previously mentioned the Pauli exclusion principle that prevents two fermionic matter particles from occupying the same quantum state. Such requirement that additional nucleons and electrons must stay separate from each other gives rise to the entire periodic table of elements. Indeed, it is the differing number of electrons in the outer orbits of atoms caused by this exclusion that manifests the wonderfully diverse characteristics and behaviors of different elements.

Any number of bosons of the same type though can inhabit the same state—for example, when enormous numbers of bosonic photons are generated to create a laser beam in which the quantum state of every photon is the same. The behavior of bosons is exquisitely perfect for the forces that weave our Universe together: light, for example, can embed vast amounts of information—one of a hologram's most important attributes.

Using musical analogies and associated terms such as *harmonic, resonance,* and *coherence* to help guide further understanding of why fundamental particles have their specific attributes isn't arbitrary—for, as we'll see, the emergence of the cosmic hologram and its holographic signatures that universally pervade space-time and at all scales of existence embody these traits. The analogy with specific notes also offers another rationale for the quantization of energy-matter in that it enables the incredible diversity that forms the cosmic symphony of our perfect Universe to express and evolve itself.

DARK STUFF

In the aftermath of the discovery of the Higgs boson, the hunt is also very much on for determining the makeup of dark matter and dark energy, which, as we've seen, together currently appear to make up a whopping 95 percent or so of all the energy and matter of our Universe. Dark matter is not only crucial in enabling galaxies to form and remain structurally stable, but it also appears to play a key role in the so-called large-scale structure of our Universe that has been progressively discovered by astronomers since the mid-1980s, as they observe galaxies distributed across the entirety of space in a cosmic web of strands and huge clusters arrayed around enormous voids.

In late 2014 observations by a European team at the Very Large Telescope (VLT) in Chile revealed even more astonishing alignments in sampling a large number of quasars: galaxies with incredibly active central supermassive black holes. Such enormous black holes, often hundreds of millions of times and, in some extreme cases, many billions of times, the mass of our Sun are thought to lie at the centers of most if not all galaxies.

The team not only discovered the alignment with the filigree of the cosmic web within which the galaxies are embedded but that, amazingly, the rotation axes of their central black holes are parallel with each other over billions of light-years, like a gorgeous cosmic string of pearls.[1]

As we've previously noted, the best candidate for dark matter is made up from some form of so-called WIMPs, or weakly interacting massive particles, and astronomers are trying to detect their presence from the gravitational effects they have on visible energy and matter. One way of doing this is by analyzing the cosmic equivalent of car crashes, when galactic clusters, some of the largest configurations in our Universe, smash into each other. One such clash involving the aptly named Bullet Cluster appears, from visible light and X-ray observations, to reveal its shadowy presence.

As dark matter is deemed to be the material underpinning whose gravitational effects hold galaxies together, in such smashes astronomers expect the visible material of the galaxies to be pulled along by the dark matter. In contrast, when intergalactic gas clouds slam into each other their visible matter slows down and lags behind their invisible dark-matter component.

While the Bullet Cluster, among others so far reviewed, is in line with these predictions, another cluster clash first studied in 2007 caused astronomers to reconsider their expectations. The results for Abell 520, an enormous merging galactic cluster about 2.4 billion light-years away, were doubted when first analyzed. Initial analysis indicated that the core of the system is abundant in hot gas and dark matter, which was identified through observations of a phenomenon known as gravitational lensing in which massive objects cause light to gravitationally bend around them rather like a lens does. However, the puzzle was that the dark-matter core seemed to be without any of the luminous galaxies that the researchers would have expected to also find in the same location.

The research team suggested a number of possible reasons for this apparent anomaly, including a more complex nature for dark matter. Later and still more sensitive analysis from the Hubble telescope has, however, been able to reveal that Abell 520 underwent a series of more complicated interactions than the others so far investigated, to the extent that it's now generally known as the Train Wreck Cluster. Such careful study has revealed the consistent behavior of dark matter and also appears to keep it in line with it being some kind of WIMP and in the simplest form such a particle can take.

The Higgs boson, though, is definitely not a candidate for such a WIMP, for the simple reason that it's far too unstable. Dark matter acts as the material bedrock to universal structure and as such must be very stable. While existing from the early epoch when the Higgs field "froze" at its lowest energy level, it may be, however, that dark matter WIMPs may be daughter products of the Higgs boson in a decay process of the entire field.

Another way of detecting the presence of dark matter is looking for the high-energy gamma rays caused when two dark matter particles collide with and annihilate each other, releasing these high-energy photons. There are thought to be three ways these can be produced, and in late 2014 a team from the University of California–Irvine studying the center of our Milky Way Galaxy found evidence of all three.[2] This and other research is suggesting a high concentration of dark matter there and indeed at the centers of other galaxies.

In early 2015 another group at the Harvard-Smithsonian Center for Astrophysics reported their findings from a review of elliptical galaxies of a direct link between the amount of dark matter in a galaxy and the size of its central supermassive black hole.[3] This not only provides another piece of evidence that dark matter molds galactic structure, but in my view it suggests something even more significant. While controversial, it leads me to ask: What if such central black holes, rather than formed from the collapse of millions of stellar black holes made up of visible matter, are actually formed from dark matter concentrations?

Whatever the nature of dark matter, the search to determine it is going into full throttle. A plethora of Earth-based experiments denoted by acronyms such as CRESST, CoGENT, DAMA/LIBRA, LUX, and XENON, the LHC at CERN, satellite-based studies such as DAMPE and PAMELA, and even the correlation of a network of thirty GPS atomic clocks to check for potential dark-matter anomalies in the fabric of space-time itself, are all racing to identify this key ingredient of our Universe, whatever that might turn out to be.

As of summer 2015, the NASA website reflected the consensus regarding dark energy when it noted, "More is unknown than known" about it. Whatever it is, dark energy, through its energy density and its interaction with other energy-matter, has an enormously important role throughout the evolutionary life cycle of our Universe.

Without it, there would be no energy of expansion to offset the attractive forces that bind matter, both visible and dark, into universal structures. Literally it's needed for space itself to expand—which, as we've seen, is the requisite for the cosmic hologram to express evermore informational and thus evolutionary complexity throughout space-time.

An intriguing clue to the nature of both dark matter and dark energy came in late 2014 with an analysis of earlier observations of CMB data from the Planck satellite that appear to show that the rate of formation of large-scale structures in our Universe is slowing down. One reason for this would be if somehow dark matter itself eventually and gradually decays into something else. While any decay into visible matter and energy is problematic for various reasons, a combined team from the Universities of Rome in Italy and Portsmouth in the UK came up with a possible solution: that dark matter slowly decays into dark energy.[4] Their analysis is tantalizing because it shows that the time frame for the increasing pace of such decay correlates with when dark energy began to predominate and the expansion of our Universe began to accelerate.

One critical aspect to determining whether such conversion of dark matter to dark energy can occur is to understand how the gravitationally attractive dark matter can somehow convert into gravitationally repulsive dark energy. And because both only interact with visible energy-matter through gravity, comprehending the nature of gravity at a deeper level may well lead to consequential insights regarding both these shadow aspects of the reality of our Universe.

VIRTUALLY THERE?

Before we discuss gravity though, we need to take a look at an assertion that's one of the most contentious in all physics: the energetic nature of the vacuum of space. I'd quite like to have avoided this topic, but it's something of an elephant in the room for physicists and cosmologists. So like all true scientists, if we're to understand it, we need to go where the evidence leads—wherever that is and whatever theoretical apple carts we upturn to get there.

Basically, the problem is that quantum field theory (QFT) appears to predict a vacuum energy density pervading all of space (sometimes called zero-point energy, or ZPE) that disagrees from actually measured values by some 10^{120} orders of magnitude (yes, you read that correctly). This humongous difference, sometimes called the vacuum catastrophe, has also been referred to as the worst theoretical prediction in the history of physics.

The issue, though, is that not only is this a severe and as yet unresolved embarrassment for physicists but that the misplaced premise of a vastly energetic vacuum is one that's fueled massive speculation, both in terms of an unknown energy flooding our Universe and in terms of the search for so-called free (or above unity) energy technologies that violate the first and second laws of information.

By another name, the real ZPE is dark energy in its most likely form of a cosmological constant that's causing the expansion of space to accelerate, which when included with all other forms of energy-matter in our Universe, both visible and dark, as we've already seen, adds up to zero throughout its lifetime.

So the enormous scale of the ZPE purported by QFT makes no sense whatsoever in actuality. And it's just as well, because if the ZPE were more energetic than it actually is, our Universe would have been ripped apart almost at the moment it began. Going back to the science for additional clarity, QFT in fact only "predicts" such a huge vacuum energy if two key assumptions are made that are being shown to be more and more unlikely.

The first presumption is that quantization of energy-matter applies down to the Planck scale. However, the developing perspective is that it's rather the pixelation of space-time at that scale that is revealing the holographic nature and informational content of our Universe.

The second is that somehow vacuum energy has some form of gravitational effects. Whether or not this may prove to be the case, a supposedly enormous value for the ZPE would cause it to then interact with the other energy-matter of our Universe in equivalently major ways that just aren't supported by observation.

As a means of understanding its underlying interactions, QFT also presumes the universal and ever-present existence of temporary or so-called virtual particles that are continually created and then annihilated in particle-antiparticle pairs whose interactions with "real" particles contribute to the nonsensical 10^{120}-out value for the ZPE. While their participation is well accounted for in many subatomic processes, describing such putative disturbances as both virtual and as particles gives many physicists the heebie-jeebies, as their transitory existence is far more nuanced.

The vacuum catastrophe then arises because, as already mentioned above, QFT presumes quantum effects at the Planck scale and also extends the presence of virtual particles beyond their involvement in subatomic interactions to pervading all of space, dramatically increasing their overall energy contribution to the vacuum.

We've seen that the first presumption is wrong and physicists are coming to reassess the second. The purported universal presence of such virtual particles, which leads to the astronomically (literally!) high and incorrect value for the ZPE, has also been presumed by supposing that they account for a number of phenomena that may instead have other explanations.

One that is often cited is the Casimir effect. This occurs when two metal plates placed parallel and very close together in a vacuum are then pushed closer together by some force. Rather than caused by ZPE, an alternative cause, as proposed by particle physicist Robert Jaffe of MIT in 2005, could be the relativistic interactions of the short-range electrostatic forces between electric charges and currents known as van der Waal's forces.[5]

In 2010 though, Stanley Brodsky and his team at the SLAC National Accelerator Laboratory in California proposed a revision of the whole notion of the universal pervasiveness of virtual particles by revising a branch of QFT called QCD, short for quantum chromodynamics.[6] QCD theoretically posits that a sea of quarks and gluons, the particles that make up protons and neutrons, pervades all of space, continually and virtually winking in and out of existence.

Instead Brodsky and his colleagues suggested that such behavior is confined to the interior structures of subatomic particles, which if correct would reduce the vacuum catastrophe by a helpful 10^{45} orders of magnitude. There's still an enormous way to go before the "worst prediction of science" is sorted out. But when it is, it's likely that the emerging understanding of the cosmic hologram will offer a major signpost to its resolution.

UNIVERSAL LAW OF ATTRACTION

Having discussed three of the four fundamental forces of our perfect Universe it is indeed time for us to come to grips with what is perhaps the most enigmatic of the four: gravity.

Einstein's 1905 special theory of relativity, as we've seen, despite ushering in a radically new understanding of space-time and the equivalence of energy and matter, only considered the behavior of objects at rest or traveling at a constant velocity. A further decade of intense thought led him, on June 2, 1915, to stand at a lectern in the large lecture hall of the Treptow Observatory in Berlin. Standing on that exact spot nearly a century afterward, I could almost hear his voice as he announced his theory of general relativity for the first time in public. In generalizing relativistic insights to include accelerating objects, and realizing that acceleration and gravitational effects are identical, he ushered in a further revolution by revealing how energy-matter dynamically warps space-time itself.

The day after his lecture, the local newspaper reported that his audience was "relatively large"—probably no pun intended. Given that his general theory wasn't yet fully completed, it's remarkable that he chose to share his insights in this first public talk before going on to do so later in the year with his scientific colleagues.

A common analogy for how a massive body curves space-time is to consider a large trampoline. When a child jumps on it, the mat is only distended by a small amount, but when a heavier adult does, the imprint is much larger. So in the so-called gravity well of our Solar System, an incoming object such as a comet can be caught up, usually by the huge gravitational attraction of our Sun but occasionally by one of the planets, which happened spectacularly to the comet Shoemaker-Levy 9 when it was caught by Jupiter's gravity and slammed into the planet in July 1994.

The consequence of such gravitational warping is the phenomenon of free fall where an object literally falls freely when there's no force acting on it and does so along a path that reflects the local curvature of space-time. So an object moving in deep, flat, space follows a straight line. Conversely, an astronaut in a stable orbit around Earth is effectively falling along an ever-circular path due to the space-time curvature caused by Earth's gravity. And someone standing on the ground is trying to fall toward the center of the Earth but prevented by the much greater repulsive forces between the bottom of her feet and the ground, as a consequence of the Pauli exclusion principle.

Rather than a fundamental force, general relativity thus views gravity as a *consequence* of energy-matter warping the geometry of space-time: a very different perspective from how EM and the strong and weak nuclear forces manifest and are mediated through their quantized bosons.

It's this difference that forms the seismic fault line between relativity and quantum theory. Much research over many years unsuccessfully tried to resolve the impasse through the hypothesis of a quantized gravitational boson called a graviton. From its side, relativity theory predicts that high gravity events such as supernova explosions should generate gravitational waves as ripples in the fabric of space-time itself.

However, the detection of gravitational waves proved extremely elusive, although study of the Hulse-Taylor binary system, where two neutron stars orbit a common center of mass, showed orbital changes consistent with loss of energy through gravitational waves as predicted by general relativity. Detecting these waves remained challenging until February 2016, when researchers announced a historic breakthrough using measurements from the Advanced LIGO (Laser Interferometer Gravitational-wave Observatory) equipment, which comprises two parallel arrays located nearly three thousand kilometers (about two thousand miles) apart in Louisiana and Washington.

From the initial vastly energetic merger between two huge black holes some 1.3 billion light-years away, they were able to record the generation of gravitational waves, enormously diluted in strength by the time they've reached Earth.[7] While their extraordinary technological achievement confirms Einstein's final prediction about the nature of space-time, it offers no insight into the possible existence or otherwise of an associated quantized particle, the graviton posited by quantum physicists.

In the coming years, even more advanced Earth- and space-based experiments, such as LISA (Laser Interferometer Space Antenna) Pathfinder, designed to be sufficiently sensitive to pick up even weaker signals, are coming into operation. Unlike the limitations of electromagnetic radiation that prevents us from probing space-time beyond the threshold of 380,000 years from the beginning when space became transparent to light, using residual gravitational waves offers the potential to peer even further back into the earliest epoch of our Universe.

Exciting!

HOLOGRAPHIC AND ENTROPIC GRAVITY

One of the intriguing implications of the cosmic hologram are the considerations it raises concerning gravity. Treating gravity as an emergent consequence of the informational and holographic structure of space-time appears to offer a more literal interpretation of general relativity than considering it as a fundamental force and might account for why physicists have been as yet unable to detect gravitons, gravity's hypothesized force-carrier particle. While very much a work in progress, such an approach may also be able to explain why gravity is so much weaker than the three fundamental forces.

This viewpoint, that both gravity and acceleration in general may be a consequence of the informationally entropic nature of space-time, goes back to the early work on black-hole informational entropy in the 1970s. In 1995, Ted Jacobson in the United States was then able to show mathematically that combining entropic considerations with the equivalence of gravity and acceleration leads to Einstein's gravitational equations.[8] Later, Indian cosmologist Thanu Padmanabhan also explored the increasingly clear link between gravity and entropy. Gravity as an emergent entropic phenomenon, however, really became a hot topic of debate among the physics community around 2010 when Dutch theoretical physicist Erik Verlinde published a paper on the subject.[9] In it he described gravity on a universal scale as the consequence of the informational entropy associated with the positions in space-time of massive bodies.

After a series of modifications to the idea by Verlinde himself and others, in 2012 Tower Wang, a physicist then at the East China University in Shanghai, unified a number of such modified approaches showing them, while constraining certain aspects, to be consistent with Einstein's gravity.[10]

From a different perspective, a number of apparently diverse theories of quantum gravity are also converging on a view that at the Planck scale all quantized fields and particles seem to behave as though space is one-dimensional and when combined with time form a two-dimensional holographic space-time.

Reducing dimensions first appeared through this approach when, in 2005, Renate Loll and her colleagues then at Utrecht University ran some computer simulations to test their idea of what they call causal dynamic triangulation (CDT), which describes how particles move apart from each other. At the Planck scale the simulations showed that the particles behave as though moving in such a two-dimensional space-time.[11]

Other quantum gravity theories such as loop quantum gravity (LQG) and so-called Horava gravity proposed by Czech theorist Petr Horava in 2009 also result in particles moving as though in a two-dimensional space-time.[12] String theory too at high Planck-scale temperatures behaves entropically as though space-time has two dimensions.

They all seem to be pointing in the same direction: to the holographic nature of space-time and gravity as an emergent entropic phenomenon.

While it's still early days, attempts at restating other fundamental forces as being entropically based are also gathering pace. In 2010, Peter Freund at the University of Chicago extended Verlinde's hypothesis of entropic gravity to a proposal that all energy-matter force fields can be described entropically.[13] And in 2011, Zhe Chang, Ming-Hua Li, and Xin Li of the Chinese Academy of Sciences obtained an apparently good fit with observational data to suggest a possible unification of dark matter and dark energy within a modified entropic force model.[14]

Time will tell which if any of these are proved correct, but the appearance of these unexpected commonalities strongly suggests that physicists are homing in to reveal something fundamental and enormously significant about the informational and holographic nature of our Universe.

5
Recipe

A way of combining things that will produce a particular result . . .

The recipe for perpetual ignorance is: be satisfied with your opinions and content with your knowledge.

Elbert Hubbard, philosopher and founder of the Roycroft artisan community, New York

In following the instructions encoded from the first moment of space-time and with the essential conditions and cosmic ingredients we've already discovered, without the exact recipe for combining these informational components, our Universe would still have been unable to exist and evolve in the way it has.

The best evidence we yet have strongly suggests that our Universe came into being perfectly balanced. The continued preservation of some of its innate symmetries, as we'll see, is crucial to the laws of physics being unchanging and the conservation of certain key properties.

A number of other initial symmetries, however, were inherently unstable and were very quickly broken. Yet, such symmetry breaking is also vital for the formation and evolution of our perfect Universe, as it resulted in the most energy efficient and thus stable non-symmetry.

In addition to the precise makeup of symmetries and the asymmetries of their breaking being essential aspects of the recipe for our perfect Universe, the other critical requirements, as we'll see, are the extraordinarily specific fine-tunings of energy-matter and interactions that pervade space-time.

SYMMETRY

In everyday terms, the notion of symmetry embodies a sense of harmony and balance. We appreciate symmetries from the frozen music of gothic architecture to the beauty of flowers and the exquisite variations of snowflakes. Too much symmetry though, as evidenced by research polls, views such apparent perfection in human facial features as apparently denoting blandness.

In mathematical terms, however, symmetry is considered in a related but much more specific way, and in physics the concept is used to describe the invariance of certain aspects of something—for example, when moved (or in physics-speak, undergoing translation) either linearly or rotationally in space or through time.

So, for instance, if two identical experiments are set up, one on Earth and the other on a planet on the far side of our Milky Way, the outcomes of both experiments would also be identical. In other words, the same laws of physics would apply: they are invariant.

Indeed if both experiments were repeated when Earth or the far distant planet had partially rotated on their axes and so in effect the experiments had undergone rotational translation, the results would still be the same. Finally, if the two experiments, instead of being separated by space, were undertaken at separate times—whether one was somehow carried out in the extreme conditions in the earliest moments of our Universe and the other on Earth today or even on another planet in the far distant future—yet again exactly the same laws of physics would apply, and the outcomes would be alike.

The invariance of space-time and consequently the laws of physics under such linear, rotational, and temporal translations is an essential symmetry for the existence of our Universe—inviolate and unbroken from its first moment until its last.

You may have had a "So what? Of course they're invariant," response to this revelation, but physicists tend to be a distrustful lot and want proof and theoretical reasons to underpin why this should be the case. It took a German mathematician, Amelie (Emmy) Noether, to supply the answer and provide physics with one of its most powerful concepts and tools for revealing the underlying order and balance of the physical world.

Emmy Noether may be the most important yet hardly known mathematical genius of the twentieth century. Born in 1882 at a time when women were virtually banned from having academic careers, after managing to obtain a doctorate, she could then only find unpaid work as a researcher for the next eight years. In 1915 though, Einstein's just published general relativity theory had pointed to profound symmetries in the relationship between space, time, and gravity, insights that, however, needed further work to fully expose.

To help reveal them, and appreciating Noether's gifts, the eminent mathematicians David Hilbert and Felix Klein invited her to join them at the University of Göttingen in Copenhagen. Even then, and after bending regulations to allow her entry at all, for some time she was only able to work on an unpaid basis. Eventually, the rules were relaxed a little, enabling her to be paid a small stipend, but still a pittance compared to what she was really worth.

Hilbert wrote later in his memoirs that he had tried to obtain a better position for her, as he was "ashamed to occupy such a preferred position beside her who I knew to be my superior as a mathematician in many respects."

Noether's huge contribution was first to mathematically prove a correspondence between symmetry and invariance. She thereby discovered an even more important connection when she was able to show that for every universal symmetry, there's a universal quantity that is conserved—a conclusion summarized in the theorem now named after her.

While the mathematical proof is rigorous and sophisticated, its upshot is the following: the translational invariance of space on a linear basis leads to the conservation of momentum and on a rotational basis to the conservation of angular momentum. The translational invariance of time leads to the conservation of energy.

Their correspondences, uncovered by Noether's theorem, also show why one of the most famous implications of the complementarity of wave-particle phenomena arises: the Heisenberg uncertainty principle. As we saw earlier, this principle describing how the ability to measure related pairs of physical qualities—where the more exact one measure is, the less exact it is for its complementary partner—is a limit imposed by our Universe itself at the level of the Planck scale.

Noether succeeded in revealing mathematically that it's the fundamental connection between symmetry and the translational invariance of space-time and consequential conservation of energy and momentum that links the attributes of Heisenberg's uncertainty principle together. So the momentum of an entity and its position in space, its angular momentum and its rotational position in space, and its energy and the time of its measurement are all, as shown by Noether, related at the most fundamental level of physical reality.

It's this primordial "real-ationship" that enables the information of the cosmic hologram to be expressed as the energy-matter of space-time. In doing so, in a subtle yet key way, it also shows that information, when embedded in such conserved qualities, must itself be conserved.

Noether's work focused on so-called continuous symmetries where the symmetry is maintained under uninterrupted and continuing translation. So, for instance, if we had the technology we could relocate our experiments through space and time in a continuous series of movements, and the symmetries, invariances, and conservations Noether uncovered would prevail. While continuous symmetry is crucial to the invariance of the laws of physics throughout space-time, our Universe also utilizes a second type of so-called discrete symmetry that embodies an either/or balance in certain traits of physical systems.

These features, such as electric charges, which can be either positive or negative, are key components of energy and matter and how they behave and interact. For example, the discrete symmetry of such charges means that for any and every particle interaction and energetic process, positive and negative charges must be created and destroyed in perfect balance. This is played out on every scale from the tiniest to the entirety of our Universe, which must be electrically neutral to exist at all.

Our Universe is replete with such discrete symmetries, where attributes are swapped or where the symmetrical partners exactly mirror each other. Indeed, they play out the on/off reflections whose most simple representations are the digital bits of information that pervade and make manifest the physical realm. In doing so, just as for bits, they embody the simplest means of denoting a fundamental duality differentiable from the whole. Yet from them, just as for informational bits, a vast array of diversity can emerge and be expressed.

The study of particle physics has revealed further symmetries in addition to the universal (or *global* in physics terms) ones we've already discussed and that act on local (or gauge) scales. Throughout the twentieth century, developments in physics progressively came to describe physical theories in terms of fields of influence. While we're familiar with the idea of electromagnetic and gravitational fields, all quantum and relativistic phenomena are now described generally in these terms.

A key aspect of such field theories is that the fields themselves can't be directly measured, that we can measure only certain observable characteristics of them such as their energies or charges. Fields can then be understood in terms of such observables whose interactions are constrained by local or gauge symmetries where the transformations differ from point to point within space-time, yet where the underlying field embeds universal symmetry.

An analogy for such gauge symmetry would be to consider a recipe for baking a cake where the recipe represents the underlying field and the flour it contains is an observable aspect. In the UK the amount of flour required for a cake would be weighed in ounces, which we'll assume is the universal unit of measure (I know, I'm biased). The same (universal) recipe in Germany, however, would measure the same amount of flour in grams. And in the United States the flour would be measured in cups. All three measurements apply to the same weight of flour within the same universal recipe, but on a gauge level, the German and US measures have a different conversion rate to the universal unit, assumed to be that used in the UK.

Wherever it's measured, the same amount of flour is weighed; it's invariant and represents the universal symmetry within the recipe. However, the observable aspects of that symmetry differ from country to country owing to the units, the ratio of the scaling from the universal to the local level.

Relating such gauge symmetries to deeper universal symmetries has been an enormously powerful tool for uncovering more profound truths about fundamental physical forces. For instance, as we've seen, quantum field theory describes the interactions of forces in terms of mediating particles called bosons (such as photons and gluons). We can now appreciate that their behavior is determined by the specific form of localized gauge transformations they take in their interactions, which is why they're also commonly referred to as gauge bosons.

In the late 1940s, the understanding of gauge symmetries led to the integration of special relativity and electromagnetism into what's called, with typical brevity by physicists, quantum electrodynamics, or more spiffily, QED. Given that folks of my era originally knew this acronym as the Latin *quod erat demonstrandum,* meaning "or so it is proved," it's not such a bad name after all for the proof of such a foundational ingredient in the recipe for our perfect Universe.

Since the discovery of QED, which American theoretical physicist Richard Feynman called "the jewel of physics," it's been the model for all subsequent field theories to date. And with its basis in the simplicity of deep symmetries, QED has caused the search for these to be the lodestone for further breakthrough discoveries.

BREAKING SYMMETRIES

While the fundamental symmetries of our Universe remain inviolate through its entire lifetime, some are directly manifested on an ongoing basis, whereas others are "broken" as a consequence of its evolution. As physicists, though, have progressively realized, it's by understanding the underlying symmetries from their apparent "broken" phenomena that offers a deeper insight as to what's really going on.

The three fundamental interactions of our Universe, the strong and weak nuclear forces and electromagnetism, all embody gauge symmetries but have different strengths, or coupling constants, and so behave very differently. A central quest in physics has been to try to integrate them in a grand unified theory, or GUT. The search has focused on trying to unify their individual gauge symmetries into a single, larger symmetry at extremely high energy levels and with a single coupling constant.

A major breakthrough in the search for a GUT occurred in 1968 when theoretical physicists Sheldon Glashow, Abdus Salam, and Steven Weinberg showed that electromagnetism and the weak nuclear forces are different aspects of a combined electroweak force. A decade later this was experimentally confirmed at the high energies required for these two apparently separate forces to reveal their deeper unification.

For the electroweak force to possibly combine with the strong nuclear force, though, requires even more extreme energies. The only time when energies have been powerful enough for such a grand unification epoch was in the earliest moments of space-time. Then, as our Universe expanded and its temperature fell, a process of symmetry breaking and phase transitions would have led to the appearance of these three diverse states of the same greater symmetry, rather like gaseous water molecules cooling into water vapor, then liquefying into water, and then cooling into solid ice.

The scientific jury is still out with regard to how such unification would actually work and what fundamental symmetry would prevail. Current experiments on Earth fall way short of the energies needed to provide direct evidence to back up any GUT or offer insights into its workings. And attempts to indirectly observe possible ramifications of a GUT have also failed.

Still hewing to the principle of "as simple as it can be but no simpler," another concern is where theoretical attempts of explication increase rather than reduce complications in terms of adding extra fields and particles. Exemplifying this is one of the leading current contenders: supersymmetry, or SUSY for short. This approach doubles the number of universal particles by postulating a supersymmetric and much more massive partner particle for every known fermion and boson. If supersymmetric particles were discovered and added to those in the standard model, the three forces could have the same strength at GUT energies, the linkage between the two types of fundamental particles would be revealed, and the lightest proposed supersymmetric particle could be a significant candidate for being the WIMP of dark matter.

SUSY may yet be shown to be an as-simple-as-possible explanatory framework, and with such ambitions, it's no wonder that its proponents are holding on, despite its being problematic in that no evidence for such supersymmetric particles has yet been directly or indirectly observed, and with the predicted window of energies for them being progressively closed by experiment.

Another leading contender is the so-called S0(10) GUT whose notation refers to a specific type of mathematical group symmetry and predicts a very massive, but as yet unobserved, type of neutrino—which is also a candidate for a dark-matter WIMP.

Anyway, we shall see.

Unlike the continuing search for a convincing GUT, which has so far taken more than forty years, the equally long hunt for another deep symmetry and its breaking was at least partially resolved when in 2012 the Higgs mechanism was finally validated, as we saw earlier, with the discovery of the Higgs boson, its force carrier.

The Higgs field underwent spontaneous symmetry breaking moments after the first moments of the big breath, when the temperature of space-time fell to a critical threshold equating to the lowest energy level of the field and that triggered the mechanism providing elementary particles with their mass. This also can be likened to a phase transition, such as the conversion of water to ice.

The investigation of every example of symmetry breaking throughout space-time has, to date, shown that all are essential. None is superfluous, and each can be identified with a vital attribute that enables our perfect Universe not only to exist but also to evolve its wonderful complexity from the most simple of recipes.

TIME, AGAIN

By now, I hope you've also come to the same conclusion as Nobel laureate Phil Anderson, who's maintained that it's only slightly overstating the case to say that all physics is the study of symmetry. But before we move on, now holding in our heads the signpost of symmetry alongside that of simplicity in our continuing quest for understanding, there's something we need to consider further: it's time to talk again about time.

Quantum physicists don't like to talk about time, as the laws of physics at the microscopic level are indifferent to it; a quantum-scale process going forward in time behaves (almost always) the same as if time were reversed. Yet as we all know and experience, at macroscales the flow of time is a unidirectional arrow.

To understand what's going on and square the circle of this apparent paradox, we need to begin by looking at another fundamental symmetry, the partial breaking of whose components is necessary for certain key physical processes to occur. Known by its acronym of CPT symmetry, it's a mainstay in quantum field theory. CPT symmetry states that our actual Universe is unable to be distinguished from one in which three discretely symmetrical attributes of a particle—C, P, and T—are simultaneously swapped for their symmetric counterparts.

The *C* in CPT stands for where a particle is replaced by its antiparticle—where the only difference is whether its charge is positive or negative, such as the antielectron or positron, which has a positive electric charge in contrast to the negative charge of the electron. Equivalently, the *P* in CPT is a switch where things look as their reflection in a mirror. Finally, the *T* in CPT signifies a reversal in the direction of time.

Experimentally, violations in all three—C, P, and most recently T—symmetries have been observed. Indeed, the breaking of combined CP symmetry in a process involving the weak force revealed a slight but crucial imbalance showing that matter and antimatter particles decay at different rates. While created, as we've seen, in perfect balance, this one in a thousand bias in decay rate may hold a vital clue as to how in the very earliest moments of our Universe a slight preponderance of matter prevailed and so allowed our Universe to survive.

When we include the T, or time, aspect in combined CPT symmetry, however, no violation has ever been observed. In other words, if we do a simultaneous triple swap of a particle for its antiparticle, reflect it in a mirror, and reverse the flow of time—voilà, we would be unable to differentiate between, for example, an electron traveling forward in time and a positron traveling backward in time.

Indeed, it turns out mathematically that the triple lockstep of CPT symmetry is mandatory for the laws of physics to be invariant throughout the space-time of our Universe.

Let's try to clarify what's happening to time at micro and macro levels and whether we can resolve their apparent contradiction.

First, in 2012 the BaBar (yes, its logo really is an elephant) experiment at the SLAC National Accelerator in the United States managed to directly measure a rare T symmetry violation, based on the different oscillation rates between particles in time-reversed scenarios.[1] While its perfect symmetry is violated, this doesn't mean, though, that time is actually reversed at these microscopic levels. It just means that this specific process runs at different speeds whether going forward or backward in time.

Significantly, it can be shown that both energy and so information expressed as energy is completely conserved during the process, regardless of the scenario's direction of time: an example of the first law of information. In other words, the process is also informationally *non*-entropic, thus not violating the second law of information. So the occasional variation in terms of the running rates of certain quantum processes is essential at the microscopic level to conserve energy-matter and to preserve the integrated symmetry of CPT.

The arrow of macroscopic time, as CalTech theoretical physicist Sean Carroll has said, is then not a consequence of the laws of physics as they apply at the microscopic scale. Rather it is the consequence of the very special and particular initial *conditions* pertaining at the first moment of the big breath: specifically, the incredibly low level of informational entropy at that first instant of our Universe.

So, it's then the second law of information that governs the one-way direction of time at the macro level where entropy and thus information embedded within such entropic processes are not conserved quantities but inevitably increase through the entire lifetime of our Universe. To exist and evolve the information encoded by our perfect Universe requires both the complementary attributes of the microscopic symmetry of time and the macroscopic asymmetry of its arrow.

FINE-TUNING

My mum baked wonderful chocolate cakes. Thankfully, she made them regularly enough that she could estimate the balancing of the ingredients within the recipe by eye and didn't need a scale. To judge by the consistently delicious results, I would guess that any variations weren't much greater than about 1 percent, or one part in 10^2.

Our perfect Universe is somewhat more precise. A while back Lee Smolin, currently a faculty member at the Perimeter Institute for Theoretical Physics in Ontario, Canada, estimated that from its beginning, if the strength of the fundamental forces of our Universe had differed only by an incredible one part in 10^{27}—an almost unimaginably minute one part in a thousand trillion trillion—our perfect Universe could not exist.

The measures of physical constants or relationships between forces need to be *exactly* what they are, otherwise our Universe would have been snuffed out before it even got going, dying at its first challenge of creating balances between energy and matter or perishing before the first stars were formed.

Ruled by the universal conservation of overall zero energy, progressive epochs are governed by the different finely tuned attributes of visible and dark energy-matter. During the big breath of the entire lifetime of our Universe, the dynamic and precise balances of these energetic forces have enabled ever-greater levels of complexity and self-aware intelligence to evolve.

In my book *The Wave,* I describe how six fundamental numbers are crucial to this exquisite harmony of our Universe. Discussion of these six numbers is also the theme of a book with that title written by UK cosmologist Sir Martin Rees.[2]

We'll now take a look at each of these vital numbers in turn in order to comprehend just how incredibly exact they've needed to be.

Strong Nuclear Force

The first number, designated by the Greek letter ε, or epsilon, is the measure of the strong nuclear force that, as we've seen, binds protons and neutrons together in atomic nuclei. In determining the efficiency of nuclear interactions ε is then also key to the alchemical process by which all ninety-eight naturally occurring elements, from the lightest, hydrogen, to the heaviest, californium, are amalgamated.

We can see how vital ε is by returning to the first few minutes of the big breath at which time the nuclei of the lightest elements—hydrogen, helium, and lithium—were forming. When hydrogen fuses to form helium, a tiny amount of the hydrogen, 0.007 of its mass, is released as energy, and it's the precise efficiency of this energy release that's denoted by ε. Thanks to ε this primordial nucleosynthesis of light elements enabled their necessary ratios to form and so go on to seed the first generation of stars and embryonic galaxies.

Then, as stars coalesce, heat up, and ignite their own hydrogen-to-helium fusion process, the efficiency of its progression as measured by ε controls its speed and so determines the stars' lifetimes. As stars age and their hydrogen fuel is used up, the value of ε then also controls the rate and subsequent synthesis of all the heavier elements that are needed to form planetary systems and biological life.

Had the numerical value been either less than 0.006 or greater than 0.008, our Sun, our Earth, and we ourselves could never have evolved.

Ratio of Electric to Gravitational Forces

The second number, N, measures the ratio between electrical forces and the vastly weaker force of gravity and is an enormous 10^{36}.

While electrical forces are crucial in holding atoms and molecules together, gravity predominates on larger scales. The reason why is because the vast majority of positive and negative atomic and molecular electric charges balance each other out, and as we've seen our Universe is overall electrically neutral. When the balance of electrical forces is disturbed, for instance, by the energetic release of electrons to form electric currents, the imbalance generally only accounts for a tiny proportion of the overall electric charges present, whereas gravity applies to all energy-matter at all times.

On the small scale of atoms and molecules though, whose masses are minute, the effect of gravity becomes negligible and electric forces predominate. If the value of N were only marginally smaller or larger, the balance of large- and small-scale forces in our Universe would have been unable to support the evolution of complexity; again, we wouldn't be here.

As we continue our journey of exploration of our perfect Universe and ultimately what it means to be human, the number N also ensures that we are poised midway in size between a molecule and a star: held in the benevolent balance of the cosmic forces that have shaped us.

Dimensions

The third number is signified by D, for dimension. Given the universal experience of space being precisely three-dimensional, its obviousness may at first elicit a "Duh!" moment from you. But please bear with me.

Not only is the choice of three-D-ness crucial to the existence of our perfect Universe and again follows the principles of symmetry and simplicity, but it's also now revealing deep truths about the holographic and informational nature of reality. The emerging understanding of the cosmic hologram as we're continuing to discover, considers the appearance of four-dimensional space-time, combining three-dimensional space with time, as emerging from the deeper reality of a two-dimensional holographic boundary of one spatial and one temporal dimension. Putting time to one side, the appearance of three-dimensional space from its one-dimensional holographic basis is the simplest way of enabling our Universe to exist.

In 2013 two teams of theorists came at the question of three-dimensional space from an informational perspective. First up were Markus Mueller at Canada's Perimeter Institute and Lluis Masanes of Bristol University in the UK who investigated the exchange of information encoded in quantum states.[3] They started only with the basic assumption of one dimension of time, a certain number of spatial dimensions and some way for information to flow through them. They then input the amounts of quantum probabilities and correlations seen in nature.

What popped out (well, not quite suddenly but at the completion of a long and complex mathematical proof) was that quantum field theory is indeed the only theory that provides the actual balance of probability and correlation, but essentially it only can do so if space is three-dimensional.

Mueller thinks this not only reveals an inextricable connection between the three-D-ness of space and the level of probability inherent in quantum theory but that the roots of both relativity and quantum theory are embedded in the way that information is exchanged in space-time.

Later in the year Borivoje Dakic, Tomasz Paterek, and Caslav Brukner at the Universities of Vienna and Singapore also looked at spatial dimensions from a deeper informational level.[4] They showed that in a universe where quantum entities interact in pairs as ours does, only in three-dimensional space can quantum theory work. This again reveals the rule of simplicity in our perfect Universe, as increasing the complexity of the interactions of quantum systems theoretically allows higher dimensional universes to exist in some form or other.

In addition to the requisite of the simplest and most symmetrical realization of quantum theory for three-dimensional space, perhaps the most important reason for three dimensions is the nature of the electromagnetic field. Pervading space, it too requires precisely three orthogonal dimensions to behave in the way it does. The electric and magnetic components of the field are at right angles to each other, with the consequential EM radiation being at right angles to both. The myriad phenomena that manifest as a result could not occur with any other than this simple and symmetrical geometry. Two spatial dimensions would be inadequate and four would be more complicated than needed.

Finally, it's just as well that D is the number that it is, as it geometrically and mathematically ensures the inverse square law of gravity that enables us to so perfectly enjoy our human experience on Earth. With two or four extended dimensions of space, either a linear or an inverse cube law would apply. In such scenarios, our wonderfully stable planetary system could never have formed.

Smoothness of Space

The next number, represented by Q, is the measure of the smoothness of space: the level of slight ripples whose frozen pattern is revealed in the cosmic microwave background, or CMB, which as we noted earlier is less than one part in a hundred thousand.

Yet, this scale of variability was perfect to ensure that as these primeval eddies of both visible and dark matter rippled through our early Universe they were sufficiently powerful to create areas of greater density. These provided seed-points that were just sufficient to then, under the influence of gravity, enable the formation of stars, galaxies, and galactic clusters, while not causing turbulence that as our Universe expanded would have prevented the stable conditions needed for stellar and galactic formation.

These primordial ripples that enabled the birth of future stars arose during an early epoch of our Universe during which it was made up of enormously hot plasma of primarily protons, neutrons, and photons that acted like a fluid. Their combating strengths of inward gravitation and outward radiation pressure produced acoustic oscillations as a series of harmonic notes.

Our perfect Universe was literally sung into existence: a Uni-verse.

This too, for me, beautifully accords with the ancient Indian tradition that teaches that all-encompassing consciousness took the form of the original vibration of our Universe through the creative sound of AUM, symbolizing the three-in-one nature of experienced reality.

Density of Energy-Matter

Not only its existence but also the future destiny of our perfect Universe is dependent on the fine-tuned precision of the fifth and sixth numbers, which relate to its innate density and rate of expansion.

The fifth number, the Greek letter *Ω,* or omega, refers to the density of energy-matter of all types within our Universe—visible and dark—and is the ratio between the actual and a critical density.

In recent years cosmologists have been able to measure Ω and found it to be the very special case of being exactly unity. Measuring the unity value of Ω is key to the acceptance of the flatness of our Universe, which, as we've seen, results in its overall attractive and repulsive energy-matter totaling to zero throughout its lifetime. Again symmetry and simplicity hold sway.

Cosmological Constant

This brings us to the sixth and final number. Again denoted by a Greek letter, this time *λ,* or lambda, it measures the cosmological constant, most likely causing the expansion of space-time and expressing itself as dark energy.

The value of λ is crucial to the future destiny of our Universe as it forms an integral component of its overall energy-matter density. It has been found to be precisely the correct strength to enable the overall density value, Ω, to total to unity.

In terms of our future this positive value for λ can also be mathematically shown to imply a finite maximum informational entropy and thus, given the complementarity of the first and second laws of information, a finite lifetime for our Universe. Its value is even more remarkable as, when compared to the offsetting force of gravity, the exquisitely evolving balance between them as space expands has played its own crucial role in enabling the evolution of complexity.

Until around five billion years ago, its subservience to gravity enabled the formation of galactic and planetary structures. But then, as spatial expansion diffused gravitational effects, it took over, and the expansion, until then slowing down, began to accelerate—much to the shock of cosmologists who discovered it in 1998.

So, since the birth of our own solar system, which occurred at the same time, the accelerated expansion of space-time has meant that the ever-increasing informational content of our Universe has been going into overdrive.

The recipe for our perfect Universe mixes symmetry and simplicity in extraordinarily fine-tuned ways. In doing so, it has and will continue to enable ever-greater levels of complexity to be encoded and embodied throughout its lifetime.

Having read the instructions, understood the conditions, assembled the ingredients, and mixed the recipe, it's now time to consider the ideal container to enable our perfect Universe to be made.

6
Container

In computing terms, any user interface component that can hold further (child) components . . .

All matter originates and exists only by virtue of a force. . . . We must assume behind this force the existence of a conscious and intelligent Mind. This Mind is the matrix of all matter.

Max Planck, originator of quantum theory and Nobel laureate in physics

So far, we've considered the informational makeup of our perfect Universe, but as many scientific pioneers have appreciated, we can't fully understand a phenomenon in and of itself but only by seeing how it relates to its wider context. In the case of the entirety of the "physical" world, cosmologists and mathematicians have long realized that to fully describe it also requires an understanding of something beyond its four-dimensional space-time appearance and from which it arises and manifests.

The inclusion of both physical and superphysical domains in attempting to describe reality is often referred to generically as being multidimensional. This term, however, has different, but as we'll go on to explore, complementary meanings in physics and the metaphysics of consciousness. We'll come to its metaphysical understanding as representing different levels of awareness and other realms of perception later, but for now let's focus on how physics and mathematics view such domains.

We'll begin by taking a further look at how the developing understanding of the cosmic hologram is describing the "container" for our Universe as being represented by a holographic boundary from which the appearance of space-time emerges. In doing so, we'll discover how the attributes of a hologram and the nature of light are perfect for the roles they play in co-creating physical reality.

We'll go on to explore how from ancient times, sages have realized that geometric relationships form the underlying patterns of physical reality and how the advent of computers have revealed that holographic fractal geometries indeed underpin and pervade the entirety of our Universe. We'll then see how the entire geometry of its curvature and shape are also crucial to its being perfect for us to have evolved.

With our accumulated understanding so far, it's then time for us to head into superphysical realms. We'll start by exploring the deeper level of actuality, where the patterns and in-formational templates of their physical real-ization reside. Without (hardly . . .) any equations in sight, we'll nonetheless discover how the cosmic language of mathematics enables anything to be transformed into waveforms and the so-called imaginary or nonphysical complex plane. As we do, we'll see revealed the in-formational and dynamic foundations of the "container" from which holographically arises the totality of our notions of four-dimensional space-time and our experience of physical reality.

LET THERE BE LIGHT

As we've seen, one of the key attributes of the cosmic hologram is that *all* energies can be expressed in terms of the frequencies of their vibrations (or the inverse, their wavelengths). Given the equivalence of energy and matter, this equally applies to particles too. Indeed, whereas energies may be considered essentially as moving waves, particles can be thought of as forms of standing waves.

The higher the frequency (or conversely the shorter the wavelength), the greater the energy carried by a specific wave; so, for example, high-frequency (short wavelength) X-rays embody more energy than low-frequency (long wavelength) radio waves.

It's ultimately through the localized and universal fields of waveform energies and their interactions that the appearance of the physical world arises.

The interaction of waves, such as the simple case of multiple ripples on the surface of a lake, is known as interference. The combined energy of waves whose peaks and troughs coincide, when they're said to be in phase, is additive, and their interference is described as being positive. Where peaks coincide with troughs, though, their energies cancel out, and their interference is deemed to be negative.

One special type of positive interference is when waves of light of identical frequency and exactly coincident phase are combined. An occurrence known as coherence, this is the underlying principle by which the intensely focused beam of a laser is produced.

The effects of such coherent focus are dramatic. A general-purpose lightbulb of a hundred-watt power can light a room. Its white light is made up of a wide range of wavelengths in the visible spectrum, and the light radiates to fill the room rather than being directed in a particular direction. A laser, however, using the same power, but utilizing coherent light of a specific frequency and focused in a tight beam, can cut though steel.

In the late nineteenth century, however, and more than a century before quantum entities and their wave-like properties were discovered, a profound insight paved the way for an encompassing description of all physical phenomena.

During the turbulent times of the French Revolution and its aftermath, the French mathematician and physicist Jean-Baptiste Fourier devised a mathematical framework that was able to deconstruct any object or energy signal, however complicated, and redefine them as combinations of simple waveforms. Using the same mathematical approach, *any* pattern of such waves can be reassembled to construct the original object or signal.

The properties of interference and coherence as embodied by light and the ability of Fourier transforms to code and decode patterns were key to the construction of the first man-made holograms and to the holographic principle. While Dennis Gabor had earlier harnessed these attributes to discover the principles of holography, it was only in 1964, following the invention of the laser, that the first hologram was constructed.

Such early holograms were produced when a single coherent laser beam, composed of light of the same frequency and phase, is split into two. One beam is directed to fall onto a light-sensitive film while the second is bounced off an object, such as an apple, effectively bathing it in a field of light and so informationally recording its complete three-dimensional appearance. The reflected light from the object is then also shone onto the film, where the overlapping beams form an interference pattern. When another laser then shines through the film, a three-dimensional holographic projection of the original object is produced.

Significantly, an aspect that is innate to the Fourier language of waveforms and the behavior of light is that the entirety of the initial object is re-created in every part of the smallest-scale pixelation of its three-dimensional hologram. So the holographic film can be cut into smaller and smaller pieces, up to the pixelation limit, and every piece will still encode the information relating to the whole.

Because information is captured from the entirety of the three-dimensional object, when the hologram is viewed from different angles, it also replicates all the three-dimensional features of the original.

While Gabor recognized that for high-definition holography, as much information as possible relating to the visible appearance of an object must be accessed, the technology of his time was inadequate to effectively do so. Developments in holography since, though, have made great strides. Recently a team of doctoral students at Tel Aviv University had the idea to dynamically alter the phase relationships between light waves to create a moving hologram. In 2014, they manufactured a prototype antenna for determining such a phase map on the nanoscale level of 10^{-12} meters.[1] For the first time, this will enable the production of high-resolution holograms capable of being projected in any direction and with the potential to allow truly dynamic holographic imagery.

Other teams around the world are racing alongside them, and in early 2015, Xuewu Xu and his colleagues at the Data Storage Institute in Singapore achieved further progress.[2] Using an array of spatial light modulators (SLMs) to alter light waves and generate three-dimensional projections, they were able to reduce the pixelation scale and so increase the overall number of pixels of information to improve the high resolution of holographic videos. However, better SLM devices with even smaller pixel size, still further-improved resolution, and faster frame rates are all needed before large-scale holographic videos can be realized.

Finally, a team led by Sriram Subramanian at the University of Bristol in the UK developed the first haptic hologram in late 2014.[3] The technology uses ultrasound projections strong enough to create tactile sensations in tandem with holographic images to enable users to have the sense of both seeing and touching the hologram.

While walk-through, moving, and interactive holographic scenarios, like a *Star Trek* holo-deck, may be achievable soon, we'll still only be able to catch a glimpse of the exquisitely simple, elegant, and profound means by which the cosmic hologram makes our perfect Universe. For, even at nanoscale pixelation levels, holograms can only encode information at a rate a hundred trillion trillion times less than that of the holographic boundary of space-time, which pixelates and encodes information at the Planck scale.

What we're continuing to discover through holographic technologies, however, are further revelations of the incredibly special properties of light. We saw earlier that being bosons, photons of light, unlike fermionic particles such as quarks and electrons, can occupy the same quantum states at the same time. This enables the amount of information embodied in the form of light to be effectively maximized. When combined with three further characteristics of the entire spectrum of electromagnetic radiation—frequency, flux, and focus—this enables huge amounts of information to flow, be stored, and be processed and in the simplest and most efficient way.

We've already seen that the higher the frequency the greater the energy or more fundamentally the information is embodied. This also applies to an increase in flux, which relates to the rate of flow or density of energy/information transfer and focus—that is, where it's concentrated in a particular direction. The optimization of all three lies at the heart of man-made holograms, and the universal qualities of light attest to the signature of the cosmic hologram.

Many physicists believe it will be proved that the electroweak interaction, which combines electromagnetism and the weak nuclear force, was also combined with the strong nuclear force in the earliest moments of our Universe and before symmetry breaking, thus forming a grand unified theory. Such integration would almost surely reveal that at its most primordial the cosmic hologram projects the entirety of physical reality as light.

While increasingly strong evidence from many scientific disciplines is pointing to the reality of the cosmic hologram, given its exceedingly tiny scale, how will it be possible to detect the signature of its holographic pixelation of space-time? Far smaller than our abilities to directly measure it, scientists need to be enormously creative at devising ways to probe space-time at this extreme Planck level.

In 2011 though, it seemed that the possibility of such pixelation might be discounted by the analysis of highly energized photons from one of the most powerful explosions in our Universe, a so-called gamma-ray burst, or GRB, emanating from a supernova. The analysis was based on a mistaken premise, but as (in scientific circles at least) it gained a lot of publicity, it needs to be debunked.

When photons travel through space, their polarization along their direction of travel is affected, increasing with the distance they've come and the energy they embody. The argument made by the experimenters who analyzed the GRB is that if space-time is smooth, the polarizations of the photons should have no preferred orientation, but that if it's grainy, as the holographic principle maintains, they should. As no bias was found, they concluded that there was no evidence of space-time pixelation.

On the contrary, if they had been able to measure any preference, for the first time ever, evidence for so-called Lorentz violation would have been found. As we've seen, the Lorentz equation exactly depicts how massive particles behave as they speed up and approach, but never quite reach, the speed of light. Photons, being massless though, always travel at the speed of light and should therefore exhibit no Lorentzian variation. So if any had been observed, we would have needed to say good-bye to special relativity and essentially all that we already know about the nature of space-time.

The primary issue is that the researchers misunderstood the basis of holographic space-time pixelation: it's based on *position* and not any single or preferred direction, and yet erroneously they were attempting to measure it along a specified line of direction.

So let's now turn to a different approach endeavoring to detect the fundamental signature of the cosmic hologram using a holographic interferometer, or holometer for short. In a man-made hologram, the innate fuzziness of the three-dimensional projection can be improved by reducing the pixelation size, leading to higher resolution of the image.

Similarly, but vastly smaller, the pixelation of space-time, according to Fermilab astrophysicist Craig Hogan, should also produce a minute fuzziness where spatial positions aren't *precisely* defined but expressed at the finite measure of the Planck scale. Extremely accurate measurements of the location of an object in two directions at the same time, according to Hogan, should then show the slight fuzziness of this fundamental pixelation. This is what the holometer aims to do.

The experiment by a team at Fermilab in Illinois led by Hogan and physicist Aaron Chou uses two L-shaped laser interferometers, each of which has two perpendicular forty-meter-long arms with detectors at their ends. If the jitter caused by Planck-scale pixelation exists, when two laser beams (split from a single source) are then run through the arms, their photons won't hit the detectors at exactly the same time.

The analysis of any such jitter is in the form of holographic "noise" that the equipment slows down to create an audio signal and that Chou, in a YouTube video, describes as listening to "the song of the Universe."[4]

The team began building prototypes in 2009, went live for a year of data collection in August 2014, and reported their initial findings in December 2015.[5] Their report, however, announced that despite their unprecedented Planck-scale level of sensitivity and equipment configuration, they'd been unable to detect any purported holographic jitter.

In response, Hogan acknowledged that if the jitter exists and is valid as a means of testing for space-time pixelation, then it's either (and in my view highly unlikely) to be at a much smaller scale than Planck or that the current setup, which assumes no universal spatial spin, may be incorrectly configured.

This indeed might be the case. One hypothesis for the shape of our Universe is as a dynamic torus with associated spatial spin. So reconfiguring the holometer to be able to detect this at the Planck scale looks to be the most likely next step.

Others too are beginning to think up innovative ways to test the nature of space-time at this incredibly tiny level. The search has really just begun.

PATTERNS OF REALITY

More than two thousand years ago sages such as Thales of Miletus, Pythagoras, Plato, Archimedes, and Euclid initiated their students philosophically, numerically, and geometrically into the harmonic nature of the Cosmos.

These initiates were called *mathematikoi,* and, armed only with the simplest of tools, they peered into the essence of reality and saw, revealed beneath its apparent diversity, archetypal and geometric patterns.

Plato, who was tutored by Socrates and himself trained Aristotle, taught in the fourth century BCE that the material world is underlain by nonphysical and abstract forms—transcendent archetypes. In the five three-dimensional solids named after him, such idealized templates were deemed to find physical expression.

Platonic philosophy and indeed his teaching academy continued to develop for centuries after his death. Not only does his influence endure, but today it is finding a new language and a rediscovered voice.

The five Platonic solids—tetrahedron, cube, octahedron, dodecahedron, and icosahedron—are the *only* three-dimensional solids possible whose sides, faces, and inner angles are the same. So, for example, the four faces of a tetrahedron are made up of four same-size equilateral triangles, the six faces of a cube by six same-size squares, and so on.

All five solids can nest within each other and with their vertices all touching a single encompassing sphere. Rotating them through different angles and seeing them from different viewpoints creates a further wealth of transformations and reveals additional harmonic relationships between these fundamental forms. Perfect indeed.

Beyond the rigorous schoolrooms of the Greek geometers though, when we look at the "real" world of clouds and rivers, mountains and coastlines, we don't see the ubiquitous presence of such impeccable, uniform, and smooth forms. We don't observe such innate regularity in physical objects and phenomena.

Or do we? If we peer beyond the appearance of diversity, as indeed the Platonists did, can we discern deeper regularity and order to our apparently roughly textured reality? As we've already discovered, the emerging understanding of the cosmic hologram from the late 1990s is beginning to transform our notions of the three-dimensional space we appear to inhabit.

However, more than a century earlier, mathematicians had already begun to deconstruct dimensionality—albeit on a basis that they thought had little relevance to the "real" world. In the early twentieth century, one of the founders of the modern study of shapes, or topology, Felix Haussdorff, studied complicated objects and how they fill the space that surrounds them. In doing so, he came to an understanding that began to extend our experience of apparently three-dimensional space and measured objects in terms of the so-called fractional dimensions that they occupied. Instead of the one-dimensionality of a line, the two-dimensionality of an area, or the three-dimensionality of a volume, Haussdorff discovered intermediate dimensions between them.

Haussdorff's work remained theoretical until Polish-born Benoit

Mandelbrot decided to see whether his predecessor's fractional dimensions could actually apply to the physical world. You can get a sense of this if you Google a map of the coastline of, say, a small island (anywhere in the world will do) and trace its one-dimensional outline on a piece of paper. If you then zoom in on the image, the coast becomes ever more convoluted the closer in you go. If you were to trace its outline at progressively zoomed in scales, you'd draw ever more intricate patterns. In essence while your trace is still a line, the increasing intricacy of the zoomed in images cover more and more of the space on the paper than a straight line would.

Unlike many mathematicians, Mandelbrot was fascinated by such real-world shapes and had a prodigious ability to perceive the geometric relationships of things and a powerful sense of their underlying order and patterns. In the 1960s and '70s, by harnessing the power of the early generations of computers to enable the analysis of huge amounts of data, he pioneered the investigation of complex and apparently chaotic systems to discover what lies beneath. After a decade-long study of apparently dissimilar phenomena such as the shapes of coastlines and fluctuations in stock-market prices, he was able to discern what no one had done before him.

Given that a straight line has one dimension and the area of a triangle, for example, has two, the outline of any rough shape in between has a fractional dimension somewhere between one and two dependent on, and increasing with, its complexity.

What Mandelbrot did was measure the shapes of natural phenomena, discovering that the British coastline has a fractional dimension of around 1.26 and that the outline of an average cloud is a bit more complex, with an equivalent dimension of 1.35. He also discovered that underlying the appearance of such complex objects are simple and self-similar geometrical patterns that replicate themselves logarithmically on smaller and larger scales. In 1975 he gave a name to these patterns of reality that he'd revealed: fractals.

Fractal geometries embrace and far extend those of classical forms. Mandelbrot's groundbreaking work, though, showed that beneath the apparent chaos and diversity of complex systems there is, as the ancient sages intuited, profound and universal harmonic order.

Ever more powerful computer analysis is discovering that such underlying fractals pervade our Universe at all scales, and crucially encode their presence not only in "natural" phenomena but throughout man-made systems. Their self-similarity and scaling up and down in logarithmic ratio is also an innate feature of holography and a further signature of the cosmic hologram.

UNIVERSAL GEOMETRY

The geometry of our entire Universe, in terms of its spatial curvature and its overall topology, also plays a crucial role in its being perfect for us.

Cosmological measurements have revealed that our Universe is spatially flat or extremely close to being so. The latest estimate from the Planck satellite mission and reported in February 2015 is of exact flatness to within an accuracy of around half of 1 percent.[6]

As we've seen too, in 2003 the WMAP analysis of the CMB showed that the tiny variations within it have a cutoff point to their wavelengths. This gives us a major clue to our Universe also being finite, as an infinite universe would instead include wavelengths of all sizes. So, given these two parameters of flatness and finiteness, what evidence, if any, do we yet have for a shape of our Universe that also adheres to these two requirements?

In 1984, Alexi Starobinski and Yakov Zeldovitch of the Landau Institute in Moscow were the first to suggest such a shape that combined flatness and finiteness—that of a torus (or doughnut to those of us who prefer our analogies edible and sweet).[7] Significantly, a torus can be considered as a rolled up flat plane, and Einstein's equations of relativity are the same for both, with the vital distinction that a torus is finite in extent, due to its rolled-up-ness, whereas a flat plane can extend infinitely.

In 1989, measurements of the CMB by NASA's Cosmic Background Explorer, or COBE, satellite led cosmologists at the University of

California–Berkeley to make the first case for what's known as a multiply-connected Universe—namely, that light can take many paths through it. The simplest geometry for such a scenario and that also allows spatial flatness is a three-dimensional torus.

Further analysis of the data from the WMAP satellite (the follow-up mission to COBE), in a 2003 study by cosmologists Max Tegmark, Angelica de Oliveira-Costa, and Andrew Hamilton, to their surprise, also revealed a greater concentration of energy along one plane of direction compared with others.[8] This alignment, jokingly dubbed the "axis of evil," provides an additional indication that our Universe may be both compact and toroidal.

More recent theoretical underpinnings to the possibility of our Universe possibly being shaped like a doughnut came in a paper published in March 2013 by a collaboration of researchers from the Universities of Southampton and Cambridge in the UK and the Nordic Institute for Theoretical Physics in Sweden and including Kostas Skenderis.[9] While the research is ongoing, the paper revealed theoretical connections between the holographic principle, flat space-time, and a toroidal shape for our Universe.

One of the cosmological predictions of a toroidal Universe, though, that arises owing to the multiple paths light can take through it is that images of distant objects should be repeated (rather like a hall of mirrors) across the sky. However, despite a number of searches, to date no such recurring pattern has been found.

Nonetheless, a 2008 analysis by physicist Frank Steiner and his colleagues at Ulm University in Germany used three different ways to compare how the temperature variations within the CMB would match up to either an infinite Universe or a finite toroidal one and concluded that a doughnut shape offers the best match.[10]

Indeed, it may be either that given the overall and increasing scale of space, light traveling along different paths may not yet have had time to create such multiple images or that the fundamentally holographic nature of space-time prevents their creation in some way.

A final thought, until we have more evidence, is that the toroidal shape, if indeed our Universe has such a topology, is actually the shape of its holographic boundary. As such, while the boundary would be multiply connected, *within* the holographic "projection" of three-dimensional space, light may only take paths that are simply connected and so would exclude any multiple images of radiating objects.

While the evidence is accumulating for the possibility that the geometry of our Universe is a dynamically evolving torus, the jury is still out. Given too that our Universe appears to be finite, the theoretical modeling of when and how it may complete its life cycle and how this could be integrated with its possible toroidal geometry, is also an open question. Finally, given that our perfect Universe began in such an incredibly special way, understanding how universes come into existence and how their births relate to the lives and deaths of previous generations is a question that we're only just beginning to ask, let alone answer.

Clearly, there's still work to do!

WHY *i*?

A while back we mentioned in passing the so-called complex plane that underlies quantum theory. It's now time for us to really begin our journey beyond space-time by encountering this nonphysical realm, and we'll start by meeting what British mathematician and philosopher Roger Penrose calls "the magic number *i*."

As a kid, I was fascinated (I know!) by the idea of the squares and square roots of numbers, especially when I understood them to represent orthogonal (or ninety-degree) dimensional shifts. The arithmetic rules that we learn at school say that regardless of whether a real number is positive or negative, its square is always positive; in other words, two negatives always multiply to form a positive.

There's a fly in this particular ointment though when we consider the simple equation $x^2 + 1 = 0$. Then $x^2 = -1$, and so $x = \sqrt{-1}$. But how can a negative number then have a square root? And what on Earth does it mean?

It's this result, $\sqrt{-1}$, the square root of minus one, that's now known as *i*. While *i* stands for "imaginary," it isn't, and actually reveals a deeper underpinning to the nature of reality that, as Penrose declares, is truly magical.

The term *i* can be adjoined to real numbers to create what's called a complex number, such as $a + ib$, where *a* and *b* are any real numbers. Within such a complex number, the terms *a* and *ib* are defined as its "real" and "imaginary" components, respectively.

The earliest tentative step toward recognizing *i* is reckoned to go back to the Greek mathematician Hero of Alexandria nearly two millennia ago. It was only in 1545, however, that Italian polymath Gerolamo Cardano in his *Ars Magna* published the first proper encounter with it.

In 1685, English mathematician John Wallis contributed a major step forward in the consideration of *i* and complex numbers in representing them graphically by creating a two-dimensional grid centered on zero. The "real" components of complex numbers are arrayed along the horizontal axis from minus quantities to the left of the zero point to positive ones on the right. The "imaginary" components then lie along the vertical axis, with negative multiples of *i* below the zero point and positive ones above. So any specific complex number, with any combinations of positive and negative "real" and "imaginary" components, can be represented as a single point on this so-called complex plane. Various arithmetical operations such as addition and multiplication of complex numbers can then be achieved through geometrical transformations of translations and rotations in the complex plane.

The use of *i* and complex numbers were still though viewed as being nothing more than a useful mathematics trick until the mid-eighteenth century when the pioneering Swiss mathematician Leonhard Euler came up with the equation since named after him: $e^{i\pi} + 1 = 0$.

Voted the most beautiful mathematical formula ever by readers of the *Mathematical Intelligencer* journal (who presumably should know), it remarkably combines *e*, Euler's number and the basis of natural logarithms that universally pervade the natural world, with *i* and π, or pi, the ratio of a circle's circumference to its diameter, with 0 and 1, the first real whole number.

Euler's equation can be restated and its square root expressed as $\sqrt{e^{i\pi}} = i$. This is extraordinary!

We've seen earlier how in our Universe, the information content of a system is entropically expressed in the form of logarithms (the second law of information) and the most fundamental geometric form is a circle. So seeing *i* embedded in such a way points the way unequivocally to the nonphysical complex plane being far more than a mathematical nicety.

Things only really heated up in our understanding of the profound significance of *i* and the complex plane as a result of the early twentieth-century scientific revolutions. The advent of quantum theory doesn't just use complex numbers as a useful tool, it actually *requires* the existence of the nonphysical complex plane that describes and plots such numbers to make sense of the deeper nature of reality.

Not only is the complex plane intrinsic to Schrödinger's equation of wave-particle behavior, but its existence pervades the entirety of quantum field theory, and, indeed, complex numbers are fundamental throughout physics and engineering.

Fourier transforms, which as we've seen apply universally and are also key to the workings of holography and provide profound insights into the cosmic hologram, require the complex plane too for their operations. We can't escape that such a nonphysical plane of existence underpins all our notions of the nature of reality. Instead of "imaginary" this is the domain in which the manifested real world is "imaged." And there's much more to come.

FRACTAL ATTRACTORS

A few years before Mandelbrot was born, two French mathematicians, Gaston Julia and Pierre Fatou, sought to map out the complex plane by undertaking a series of iterations—mathematical processes of progressive change.

During the turmoil and tragedy of World War I, the Frenchmen found in their analyses the deep ordering of attractors—naturally arising points in the complex plane that attracted surrounding points to them—and, conversely, other points that acted as repellers. The boundaries around the catchment areas of the attractors were made up of repelling points and proved to be extremely complicated—so much so that decades before the advent of computers enabled the revelation of their exquisite details, the two men never saw the incredibly beautiful fractal forms that their attractors and repellers exhibit and that are now known as Julia sets. What they discovered, though, would be instrumental for Mandelbrot's later work and how such fractal attractors apply to apparently chaotic and complex systems.

Another critical contributor to Mandelbrot's discoveries was American mathematical physicist Mitchell Feigenbaum, who also investigated the nature of iterations. He was especially interested in understanding how, when variables within systems pass certain thresholds, the entire system undergoes bifurcation, or branching.

The more sensitive a system is to a certain variable increases its responses and leads to further bifurcations with each branch itself splitting into a further two. Unless halted in some way, the rate of the continued doubling of the splitting approaches ever closer to a single point, known as the Feigenbaum point, and where the system enters a chaotic state. Vitally though, even within the chaos are areas of order that can become the fractal seeds of emergent new stability.

Feigenbaum found that the ratios between successive bifurcations converged to a specific number, 4.6692016. . . , *regardless* of whatever systems are studied. Like π, another never ending and never repeating irrational number, Feigenbaum's number is a truly universal constant of our Universe.

His understanding, however, emerged from studies concerning and solely measuring real numbers. It was only when Mandelbrot used the power of computer analysis to combine the understanding of attractors and Julia sets with the extension of Feigenbaum's insights to the complex plane that a vastly greater and even more profound picture emerged.

Mandelbrot was aware that Julia and Fatou had discovered two types of fractal Julia sets. One is made up of completely disconnected groups and the other of completely connected ones, and he wanted to understand the deeper pattern that caused the two types. He did so by iterating the simplest form of transformation that generates further complex numbers and effectively maps the complex plane onto itself.

In 1980 he decided to map the mapping by coloring in points from which arose connected Julia sets and leaving blank the points from which disconnected sets emerged. This identified all connected and disconnected Julia sets in what became a single larger set, now famous and named after Mandelbrot—for what came out of the computer was as utterly unexpected as it was utterly wonderful.

The inside and outside of the geometrical Mandelbrot set was the boundary between all connected Julia sets (inside) and all disconnected Julia sets (outside). It's the boundary of the Mandelbrot set, though, that's truly extraordinary and that marks the embedded Julia sets as being exactly on the cusp between connection and disconnection. It's here where zooming in at smaller and smaller scales reveals ever smaller and embedded Julia sets; Mandelbrot had literally discovered the fractal harmonics of creation.

What they opened the doorway to was a realization that all physical complex and chaotic systems are based on underlying nonphysical complex plane attractors and fractal patterns and how the complicated and diverse appearance of the "real" world arises from the intrinsic order and instructions of simple rules, principles, relationships, and geometrical patterns in a realm beyond space-time. These harmonics of creation, and whose attributes are those of the cosmic hologram, underpin the entirety of our Universe at all scales of its existence.

It's by understanding the in-formational content of such attractors that further insight into the premise—that, while probabilistic, no phenomenon is ultimately random—is gained.

Throughout the holographic processes that on myriad levels play throughout the whole world, individual probabilities accrue to form collective determinism. For example, the heights of individual people vary, yet they're not "random" but probabilistic. The bio-field template of a human being holds information that results in a statistical range that goes from the scale of a newborn baby to the height of the tallest person in the world. Collectively, the range is deterministic. So, if the heights of everyone living were plotted on a graph against their frequency, the number of people at that height, the overall distribution would be in the shape of a bell, whose peak corresponds to the overall average. As is the case for heights, such so-called Gaussian curves describe the normal distributions of characteristics of many types of populations.

Indeed it turns out that the distribution for Gaussian curves, which arise from a family of stable fractal attractors, also embody within any population the *maximum* amount of informational entropy for a given average and range of variance for the specific characteristic analyzed.

DEEPER DIMENSIONS STILL

Study of the even deeper nature of the complex plane, its ability to encode in-formation and what may lie beyond, is at the very leading edge of science. Just as in the early twentieth century the complex plane went from being a mathematical tool to the realization of deeper nonphysical realities, so in the early twenty-first century the probing of other nonphysical "spaces," such as phase or momentum space, hitherto deemed merely mathematical, may reveal even more revolutionary discoveries and greater perspectives on the profound nature of reality.

These may well also include discovery of the existence of higher dimensions—either tiny and curled up as proposed by M-theory or large-scale and perhaps incorporating as yet unknown characteristics of light.

Given the innate in-formational nature of physical reality, understanding how the potentialities of in-formation may both arise in such higher dimensions and then "intention-alize" into manifestation is vital. Such an understanding of the emergence of the "real" world from its "imaginary" "field of all possibilities," as my colleague Deepak Chopra has described it, is ultimately the most important of all scientific and indeed spiritual quests.

It not only has the capacity to widen our perspective of the Cosmos but also to offer transformational steps in answering our essentially human questions of "Who am I?" "Why am I here?" and "Where am I going?"

7
Perfect Outcome

Having all the required or desirable elements, qualities, or characteristics: as good as it is possible to be . . .

To live is to change, and to be perfect is to have changed often.

Blessed John Henry Newman,
theologian and academic

The emerging twenty-first-century scientific paradigm of the cosmic hologram has the transformative potential to integrate and radically expand on the scientific revolutions of the twentieth century, with their perception of the relativity of space and time and their integration into invariant space-time and the quantization and equivalence of energy and matter.

To paraphrase German mathematician Hermann Minkowski, who pioneered the geometrical and relativistic understanding of space-time, and update his insights for our own era of groundbreaking discovery, henceforth space-time by itself, and energy-matter by itself, are doomed to fade away into mere shadows, and only a kind of union of the two will preserve an independent reality.

The increasing realization is that it's information—underpinning and pervading all that we call physical reality—that offers that union, that from the dynamic in-formational patterns embedded on a holographic boundary and arising from deeper nonphysical realms, the existence and evolution of our entire Universe, including our experiences as human beings, is played out.

We're now ready to bring together the discoveries that we've explored so far to summarize how this developing understanding of the cosmic hologram is revealing how the perfect outcome for our Universe is co-created.

IN-FORMATION

Information, as discussed in chapter 1, literally in-forms all that we call physical reality, and from the innate instructions, conditions, ingredients, recipe, and container of the in-formation that make up the cosmic hologram, enables the outcome of a Universe that nurtures the evolution of complexity and ever more self-aware consciousness—makes a Universe that is perfect for us.

As physicality is being progressively shown to be immaterial and information to be physical, their congruence is leading to an increasing awareness that to understand the essential wholeness of reality requires that the principles and laws of physics be restated in informational terms.

At every scale from the most minute up to its entirety, the reality of our Universe is indeed being restated in this way, revealing itself as being constituted of holographically expressed in-formation, which is more fundamental than space-time and energy-matter. Quantum informational behavior and nonlocal connectivity, which transcend the limits of space-time, aren't restricted to the quantum level. Experimentally, nonlocality has been proved for objects as large as small diamonds, and, theoretically, it is intrinsic to the most fundamental forces of our Universe.

There's no fundamental difference between quantum and macroscopic scales. They only appear different owing to the difficulty of informationally isolating larger entities from their surroundings. This shows that our Universe is innately coherent and nonlocally unified, where everything is fundamentally interconnected and informational in nature.

Evidence is growing that at the tiniest Planck scale, space-time is pixelated; this is the level that is foundational for informational and holographic reality. Both from theory and cosmological observations there are increasingly strong indications that our Universe is finite in terms of both time and space—being born, living and eventually dying within an infinite and eternal cosmic plenum.

A finite universe, though, can only embody finite information. So there needs to be a mechanism for the essentially unlimited, continuous, and nonphysical wave functions of quantum potentials to become finitely real-ized. Quantization, with its discrete nature, is such a mechanism.

In-formation expressed as digitized bits, with one bit encoded per Planck-scale area by the holographic boundary, is then the simplest and most efficient method for its communication and processing.

INSTRUCTIONS

As explored in chapter 2 our Universe was born 13.8 billion years ago in the so-called big bang, which was neither big nor a bang. Instead, at the first and tiniest moment of space-time it embodied immaculate order, expanding not in an explosive burst but with incredibly fine-tuned precision: the big breath.

From its birth it encoded the complete information and algorithmic instructions that ensured that all laws of physics pertaining to the behavior of energy-matter and that are described by quantum theory prevail universally and so enable it to exist as a unified entity. Such encoding and coherence also empowered the creation of elementary particles and the fundamental processes and interactions that progressively gave rise to stars, galaxies, and the evolution of ever-greater complexity and diversity.

Cosmological measurements have also shown that space is geometrically flat, a very special circumstance that is both crucial for the relativity of space and time and their integration as invariant space-time. The finiteness of our Universe, its flatness and the expansion of space, also means that information expressed as energy-matter, visible and dark, is both conserved and balances exactly to zero throughout its entire lifetime.

Such conservation of information expressed as energy-matter on a universal basis is a statement of the first law of information. As such, *the first law of information is essentially also the generalized expression of quantum theory.*

In addition, the origin of our Universe, in an extraordinarily ordered state, embodied its minimum informational entropy that ever since has increased inexorably, causing the arrow of time to flow and the principle of causality within space-time to be inviolate.

The continually increasing entropic flow of information within space-time, rising to a maximum at the end of the lifetime of our Universe, has enabled the development of ever-higher levels of consciousness and self-awareness to be expressed, embodied, and experienced. Such an informationally entropic process of both space and time in their integration in space-time from a minimum to maximum state is a statement of the second law of information.

The nature of time itself can even be considered as being the accumulated flow of informational entropy, ever increasing from past to present to future. Indeed, just as *the first law of information is the generalized expression of quantum theory,* so the *second law of information is that for relativity theory.*

Restating and expanding the nineteenth-century laws of thermodynamics to twenty-first-century laws of information (or infodynamics), in providing a deeper understanding of the nature of physical reality, itself also perceives the complementarity and thus enables the integration, as noted previously, of the two hitherto unreconciled pillars of twentieth-century science: quantum and relativity theories.

The two laws of information indeed reveal that there's no need to try to crunch quantum and relativity theories together, as has been tried for the past eighty years or so, but instead to comprehend them as being complementary attributes of in-formational physical reality, the first depicting universally conserved energy-matter, the second universally entropic space-time.

The first law of information enables our Universe to exist; the second law enables it to evolve.

The study of general relativity and the informational entropy of black holes have also shown that their maximum amount of information, pixelated at the Planck scale, isn't proportional to the three-dimensional volume of space that they occupy but rather to their two-dimensional surface areas.

This can be extended to all the information embodied by the apparently four-dimensional space-time of our Universe, which may instead be considered as being encoded on a two-dimensional holographic boundary. For ever more information in the form of evolution and increasing complexity of experience and awareness to be expressed within our Universe, the total area of its two-dimensional boundary *must* expand—which as its ongoing big breath, space has and is continuing to do so. Both space and time are thus emergent phenomena of the informational cosmic hologram, and the nature of space-time is entropically expressed in informational terms by the second law.

The logical conclusion of our Universe's unfolding informationally entropic process suggests that its end will be as a maximal entropic state and complete thermal equilibrium at, or very close to, absolute zero. While as yet there's no known mechanism, this end-time may engender the release of its accumulated information, knowledge, and wisdom into the infinity of the eternal cosmic plenum.

CONDITIONS

As noted in chapter 3, simplicity, invariance, and causality are three fundamental conditions of our perfect Universe.

Universal simplicity is revealed through deeper amalgamations of diverse phenomena and laws of physics that while supporting the emergence of evolutionary complexity are essentially as simple as they can be but no simpler.

Universal invariance shows that whereas space and time are each relative to the position of an observer and woven together by the constant speed limit of light, their integration as four-dimensional space-time is invariant. So, all measurements of the same event in space-time will give the same answer to all observers, regardless of their locations.

The speed of light ensures that information is transferred at such a maximum rate *within* space-time, maintaining universal causality. Yet the phenomenon of universal nonlocality is a further revelation that our Universe as a whole is simultaneously interconnected and evolves as a coherent and unified entity.

INGREDIENTS

As discussed in chapter 4, there's only one ingredient that makes up our perfect Universe—information—expressed and conserved as energy-matter in all its interchangeable forms and relentlessly increasing through entropic processes.

Without matter having mass, everything within space-time would travel at the speed of light, which massless photons of light do, and the passage of time would stop. Fundamentally, the gaining of mass by elementary particles, imbued by the Higgs field, slows things down, enabling the entropic flow of information and the experience of time itself.

The different attributes of all elementary particles are vital for the existence and evolution of our perfect Universe and are now being understood in terms of their harmonic, resonant, and coherent behaviors—all signs of the cosmic hologram. Three fundamental interactions—electromagnetism and the strong and weak nuclear forces—may well have been integrated into a grand unified force at the extreme conditions just after the beginning of our Universe, most likely with the characteristics of light. When space expanded and energies fell below a certain threshold they underwent forms of phase transitions and separated out into the three subsequent types of interactions.

Quantum and relativity theories describe the behavior of energy-matter differently on small and large scales. Previously, theories attempting to reconcile and integrate them have tried to quantize gravity at extreme energies. As we've seen, the first and second laws of information show that instead they are complementary in their depictions of conserved energy-matter and entropic space-time.

Treating gravity, then, as an emergent consequence of the informationally entropic nature of space-time and, critically, the *acceleration* it engenders between masses, may offer a more effective means of integrating it within the emerging perspective of the cosmic hologram.

All approaches to such integration also seem to require a reduction in the number of spatial dimensions. The three we're familiar with fall to a single dimension that combines with time to form a two-dimensional space-time—another sign of the holographic foundations of physical reality.

RECIPE

The *exact* recipe for combining the informational components of our Universe is essential for it to exist and evolve in the way it has, as discussed in chapter 5.

Our Universe appears to have come into being perfectly balanced, yet inherently unstable. Continuation of some of its initial symmetries is key to the laws of physics being unchanging as well as the universal conservation of certain fundamental properties, such as energy-matter. The instability of other symmetries, though, caused them to be quickly broken, such as matter and antimatter, resulting in the most energetically efficient and stable non-symmetries; such asymmetrical processes are also vital for its ongoing existence and evolution.

In addition to the extremely special nature of the flatness of space, the types and balances of energy-matter and their interactions were incredibly finely tuned from our Universe's first moment; otherwise it would have perished before it even got going.

CONTAINER

As discussed in chapter 6, the cosmic hologram describes the "container" for our Universe as being represented by a holographic boundary informationally pixelated at the Planck scale from which the presence of space-time emerges; the search for evidence for this "container" is ongoing.

The informational and energetic attributes of the holographic principle and the nature of light are ideal for the co-creation of physical reality, enabling the maximum amount of information to be most simply and efficiently embodied. The vibrational nature of all types of energy-matter allows any object to be mathematically redefined as a combination of simple waveforms and reassembled to reconstruct the original object. Such Fourier transforms are fundamental to the holographic principle.

Following pioneering work by Benoit Mandelbrot, computer analysis has revealed that the fractional dimensions of self-similar fractal geometrical patterns holographically underpin and pervade our Universe at all scales, encoding their presence not only throughout "natural" phenomena but throughout our man-made world too. The innate relationships of geometric forms are fundamental aspects of the cosmic language of mathematics that all reality is written in.

While there remain unanswered questions, with the proven flatness of space and increasing evidence for its being finite, as we've seen, it's the geometry of a torus or doughnut that seems to be the best fit for the shape of the holographic boundary of our entire Universe.

Our perfect Universe can only be understood in terms of its manifesting from a more fundamental reality beyond its four-dimensional space-time appearance. The concept of i (the square root of -1), complex numbers, and the so-called complex plane, far more than just mathematical tools, are indeed the essential superphysical underpinning to the appearance of physical reality.

Quantum theory, Fourier transforms, and many other theories and processes that describe the nature of physical reality actually *require* the existence, beyond space-time, of the complex plane. It's here too where superphysical patterns of fractal attractors lie beneath the apparent chaos and diversity of complex physical systems, providing more evidence, as ancient sages intuited, of deeper and universal harmonic order.

Such a nonphysical plane of existence underlies all our notions of the nature of physicalized reality. In the early twentieth century the complex plane went from being a mathematical tool to the realization of a deeper nonphysical but innately in-formational reality. In the early twenty-first century, the probing of other nonphysical "spaces" hitherto deemed merely mathematical may reveal even more profound and transformational discoveries.

ENJOYING THE PERFECT OUTCOME

Now that we've recapped the information, instructions, conditions, ingredients, recipe, and container for "making" our Universe, we're ready to enjoy its perfect outcome. The fundamental attributes of the cosmic hologram that *is* the physical reality of our Universe are all innate in the information that's universally expressed as conserved energy-matter and entropically expressed as space-time.

Such in-formation makes up the exquisite order and fine-tuning of its instructions, the wonderfully elegant simplicity of its initial conditions, the amazing versatility of energy-matter that constitutes its single ingredient, the incredible exactness of the recipe that defines all its interactions and processes, and the ideal nature of its holographic and pixelated container. From its nonphysical foundations, the cosmic hologram encodes informational fractal patterns of potentiality that manifest throughout space-time at all scales of existence and that dynamically guide the templates of evolution.

For ourselves, perhaps most important is its essential nature that enables our Universe to exist and evolve as a nonlocally coherent and unified entity from its first moment until its last, empowering the advancement throughout its lifetime of ever-greater levels of complexity and the emergence of self-aware intelligences like us.

This perfect outcome is surely delicious. Bon appétit!

PART 2

◇◇◇◇◇◇◇◇

Our In-formed and Holographic Universe

8
Universal Patterns

Everything in the world embodied in ordered, self-similar, and intelligible forms . . .

Beauty is truth's smile when she beholds her face in a perfect mirror.

Rabindranath Tagore, creative artist and Nobel laureate in literature

Thanks to the exponentially increasing power of computers to analyze enormous amounts of data across broad areas of study, scientists are becoming ever-more aware that in-formational fractal patterns, at all scales of existence from the quantum level to galactic clusters and beyond, permeate our Universe.

In this and the following two chapters, we'll explore how they and complementary forms of universal connections are revealing the intrinsic simplicity and innate wholeness of our Universe and, importantly, how the same patterns are being discovered throughout both "natural" and "man-made" phenomena.

We'll also see how nonphysical fractal attractors underpin all complex phenomena in the physical world and how universally harmonic and coherent relationships among key variables in a vast range of situations incorporate the same fractal and holographic relationships and processes that are revealing the ever-present and fundamental signature of the cosmic hologram.

FRACTALS

Having introduced fractals in general and their self-similar and holographic nature, it's time for us to see just how all-pervasive these underlying patterns really are. While we'll return in more detail later to their universality throughout biological systems and our human choices and behaviors, for the moment we'll focus on how a wide range of scientific disciplines is discovering how they abound through "natural" phenomena.

Before we do though, we need to expand on my description of fractals as being self-similar. For, while indeed in many cases they are, to be precisely correct this self-similarity requires their scaling up and scaling down to be in exact ratios of their basic pattern. Where their dimensions scale up and down at different rates to each other, they're instead more generally defined as being self-affine. Self-affinity is a term that denotes the wider resemblance of the parts of a system to its whole. Self-similarity is thus a specific subset of the more generic self-affinity, which is, of course, still scale-invariant and holographic in nature.

Studies of planetary topography, for example, demonstrate both self-similar and self-affine fractals, with land surfaces tending to feature self-similar patterning and the cross sections of vertical elevations revealing self-affine structures.

An example in which self-affine fractals prevail is the behavior over time of stock markets, where the often turbulent conduct of the system also requires a further iteration of fractal patterning. The markets' volatilities entail more than one self-affine fractal to better track and predict their movements. It was Benoit Mandelbrot's pioneering work on stock-price models that led him to coin the term *multifractals* for the multiple forms needed to describe such complicated variability.

In considering the following "natural" systems, as well as later when we'll look at biological and "man-made" phenomena, we need to keep in mind that most of them have levels of variation that can be best understood in these wider contexts of self-affinity and multifractals.

Geology and geophysics are full of scale-invariant processes and occurrences that exhibit chaotic behavior whose turbulent nature masks underlying order that is invariably fractal. So fractal modeling throughout Earth sciences has many applications, from the analysis of mineral reserves to the fragmentation of tectonic plates. Throughout, many fractal structures show similarities across manifestations that appear to be very different, such as the topology of the Grand Canyon and the Idaho batholith, which in 1991 geophysicist Donald Turcotte, then at Cornell University, showed share the same fractal dimension.[1]

We've already noted that coastlines are fractal, but the drainage of river systems and the sizes of mountains within ranges are too, and even forest fires spread out along fractal boundaries.

In meteorology, not only clouds but also lightning cascades and the repeating patterns of snowflakes are fractal, and we'll see later how the still-greater complexity of entire weather patterns arises from underlying fractal attractors.

All chemical processes involve energy, and thus information flows, which are inherently fractal. The applications of fractal insights to chemistry and materials science are increasingly common as, for example, where the fractal patterns seen in corrosion phenomena are helping a better understanding of the surface structures of metals.

Throughout physics too, the ubiquitous nature of fractals is progressively showing up at all scales. In 2010, Ali Yazdani, then at Princeton University, and his colleagues discovered that these patterns exist at the scale of single atoms in a solid.[2] At the sudden transition point where a material changes from behaving as a metal to acting as an insulator, the team used a scanning tunneling microscope (STM) to directly observe the fractal clustering of electrons at the atomic scale.

In another breakthrough, in 2013, and for the first time, physicists reported experimental proof of one of the earliest fractal patterns theorized in quantum physics.[3] It's known as the Hofstadter butterfly after its discoverer, Douglas Hofstadter, who had proposed the fractal structure to describe the motion of electrons in extreme magnetic fields. He posited that as a magnetic field is increased, the quantum-energy levels of electrons confined within a crystalline lattice would progressively split, and when represented on a graph would show a fractal pattern resembling the wings of a butterfly.

At the time of his insight in the 1970s, Hofstadter was a young graduate student, and, not uniquely in the annals of physics, his doctoral advisor was distinctly unimpressed by his suggestion. For more than forty years his idea was very difficult to experimentally test. But finally using graphene as a proxy model, Hofstadter's predicted electron behavior was revealed, albeit indirectly. As Pablo Jarillo-Herrero, one of the team, noted, "We found a cocoon." As he then added, "No one doubts that there's a butterfly inside."

Another discovery of underlying fractal structure relates to the phenomenon of superconductors that have no electrical resistance and so are extremely useful in being able to conduct a large current with no loss of energy. Completely unexpectedly, when researching properties of crystalline superconductors in 2010, Antonio Bianconi and his team at the Sapienza University of Rome in Italy revealed that not only was the crystal structure fractal but also that the longer the length scales for which the fractal pattern lasted, the higher the temperature for which the superconductivity persisted.[4] So engineering such crystals to maximize their fractality could produce materials that are superconductive at even higher temperatures.

At much larger scales, astronomers and cosmologists too are discovering that fractal patterns pervade our Universe. Within our Solar System, for example, the solar wind, the turbulent stream of charged particles that our Sun constantly emits, was found in 2014 by Sandra Chapman and her colleagues at the University of Warwick in the UK to have fractal properties.[5] Saturn's rings too reveal a fractal pattern.[6]

Further afield, study of the CMB (the astronomical gift that just keeps on giving) has shown that galaxies also cluster in fractal patterns. In 2012, having undertaken the largest three-dimensional mapping so far of such large-scale clustering, Morag Scrimgeour and her team at the University of Western Australia in Perth analyzed data from the WiggleZ Dark Energy Survey.[7] Using observations by the Anglo-Australian Telescope in New South Wales the team charted some 220,000 galaxies throughout an enormous volume of space corresponding to a cube with a length of three billion light-years to a side. At a scale up to 330 million light-years across they did find fractal galactic clustering. But beyond that, galactic spacing appeared homogenous, which is in line with the model of the cosmic hologram that we've been exploring. For, while fractals, scale-invariance, and other holographic attributes are features of the manifestation of space-time, there's no need for the overall topology of our Universe itself to be fractal.

Indeed if it were so, enormous problems for its study would have ensued, not least that our notions of space-time itself and the use of general relativity to model it, which has passed every experimental test it has so far faced, would have been invalidated and would have had to have been thrown out or at least radically modified.

The largest scale fractal patterning yet discovered by Scrimgeour's team are in agreement with the maximum size to be expected in terms of gravitational organization of matter over the 13.8 billion years of our Universe's big breath. What the repetition of fractal patterning from the smallest to the largest scales shows us, however, is that the informational patterns that underlie our Universe embody the minimum information and simplest instructions at all scales to enable manifestation of the maximum diversity and the development and evolution of the greatest complexity.

The discoveries that underlying fractal patterns are universally pervasive also reveal the innate harmonic order of our Universe. But for us to continue to see what this means for our wider understanding, we first need to consider what I mean by harmonic order and how its simple rules play out in complicated ways.

HARMONIC ORDER

When I was at school many years ago, and well before the invention of pocket calculators, in my math class I learned how to use two indispensable tools: log tables and a slide rule. And while electronic calculators and computers have long since made both obsolete, how and why they worked is as pertinent—and with the understanding of the cosmic hologram and the harmonic nature of reality—even more relevant than ever. Let me explain why.

What they both enabled was for otherwise complicated sums such as multiplication, division, and calculating square roots of numbers to be dramatically simplified. They did this by using logarithms, which convert multiplication into addition and division into subtraction—much easier!

You'll probably be grateful that I'm not about to go into the technicalities of logarithms. What's key for us here though is to appreciate that they operate in the way they do because essentially they're harmonic. This is owing to their being exponents (or powers) of a base number, such as 10, where the base is multiplied by itself however many times to arrive at a required number.

To explain this numerically, we need a bit of math. The log of any number x (say 100) is the power to which the base (in this case 10) would need to be raised to equal x; in this case 2, as 100 is 10×10. So the log of 100 in base 10 is 2, or $\log_{10}(100) = 2$.

This allows the conversion of numbers to their log equivalent, and vice versa, and the harmonic mathematical properties of logs to be then used to simplify complicated calculations. For optimum ease, base 10 (or common) logs are used in general science and engineering and base 2 (or binary) logs in computing science. Base e (or natural) logarithmic relationships, though, where e is Euler's number, pervade mathematics and are especially important given the deep connection between e and the complex plane.

An essential consequence of their harmonic nature is that logarithms are rather like geometric fractals in that they are self-similar and scale up and down by factors of their numerical base. Logs are found through many universal relationships, and logarithmic spirals are key features of numerous evolutionary processes. We even see and hear the world logarithmically, because our eyes respond in this way to the brightness of visible light and our ears to the logarithmic decibel scale of the frequencies of audible sounds. Such logarithmic sensory perception enables us to see and hear across a wider range of visual and acoustic stimuli without overwhelming our senses.

Natural logarithms pervade Fourier transforms that are basic to the holographic principle. Most significant of all perhaps, they anchor the fundamental equation for informational entropy and so again reveal the nature of reality in innately harmonic, fractal, and holographic terms.

A while back we explored how nothing is ultimately random and considered a common distribution of data, such as the heights of a group of people, that overall fit into a so-called Gaussian distribution curve in the shape of a bell and where the average measure of the group correlates to the peak of the bell.

There are, though, many phenomena whose features don't correspond in this way. Some key ones obey instead a distribution based on log-based relationships. These are called power laws and incorporate logarithmic correlations between sizes and frequencies of their occurrences. Crucially, such phenomena are holographically self-similar at all scales over which they apply, and so there's no such thing as an average measure or typical event.

One example was revealed in 1989 by Lindsay McClelland and his colleagues. Studying the relationship between size and frequency of two tranches of global volcanic eruptions, the first going back two hundred years and the second the decade between 1975 and 1985, they discovered their scale-invariant nature over an extensive range.[8]

A further such manifestation is the incidence of earthquakes, which follows the Gutenberg-Richter power law for their magnitudes, where the relationship between the size of a specific quake, as measured logarithmically, has a power-law relationship with its frequency.

The logarithmic scale uses base 10 logs to measure the energy released. An increase in scale of 0.2 represents a doubling of the energy. So, a massive 9.0 event, such as the Japanese earthquake of March 11, 2011 (which I personally experienced in Tokyo), and which was the largest yet recorded there, released 64 times the energy of the still terrifying and disastrous magnitude 7.8 Nepalese earthquake of April 25, 2015.

What this harmonic power law shows is a simple yet crucial relationship that applies to *all* earthquakes of whatever scale and wherever and whenever they occur: one of twice the size occurs four times less often.

Such scale-invariant relationships between the size and frequency of power-law phenomena reveal two further key insights. First, the *energy* of *each* earthquake multiplied by its *frequency* remains *constant*. Second, while there are fewer major events than smaller ones, there's no way of predicting the time and location of specific occurrences.

The best we might hope for is an early-warning system that can measure the stress buildup in vulnerable geological strata, although even then, as a nonlinear process, the size of the stress is not necessarily indicative of the scale of any resultant quake.

We may indeed already have such a system. As suggested by British biologist Rupert Sheldrake, perhaps a more open-minded and thorough investigation into the widespread anecdotal evidence that animals and birds may, owing to their inherent sensory sensitivity, show that they do pick up subtle environmental changes presaging such events.

Sheldrake proposes a hotline or website that people could contact to report strange behavior in their animals. Using computer analysis to determine any clusters that aren't attributable to other factors that correlate to a subsequent quake and can be read in conjunction with other seismological measuring devices, such an initiative could aid improved prediction and empower appropriate action.

THE FAIREST BOND

Another exquisite example of harmonic order is what Plato considered to be the "fairest bond" in the physical world: when a line is divided into two, such that the ratio of the smaller to the larger segment is equal to that between the larger segment and the whole.

This universal relationship is known as the golden section, or phi (designated by the Greek letter φ). Its numerical properties are astounding, and so significant is its expression throughout nature that it was called by the fifteenth-century Italian mathematician Fra Luca Pacioli "the divine proportion." Indeed, Fra Luca wrote a book of the same name, illustrated with beautiful geometrical drawings by no less an artist than his friend Leonardo da Vinci. Its aesthetic beauty has been reflected in the work of numerous artists and architects and its proportion is even found in the basic shape of modern credit cards. Many profound thinkers too have been fascinated by its attributes and pervasive presence in evolutionary systems, including the World War II British code breaker and father of computer science, Alan Turing.

Mathematically, phi is termed an irrational number (like pi and Feigenbaum's number) as the digits after the decimal point in its value of 0.61803. . . continue without repetition forever. Uniquely, 1 divided by phi is equal to phi plus 1, or 1.61803. . . also referred to as phi.

Earlier, in the thirteenth century the Italian mathematician Leonardo of Pisa, more commonly known as Fibonacci, revealed other profound aspects of phi through a series of numbers that has come to bear his name. Starting with zero and then one, each successive term in the series is the sum of the previous two numbers: 0, 1, 1, 2, 3, 5, 8, 13, 21, 34, 55, 89, 144. . . , and so on, indefinitely.

Phi arises when each number of the series is then divided by the one after: 0/1, 1/1, 1/2, 2/3, 3/5, 5/8, 8/13, and so on. By then converting these ratios to decimal form and plotting them on a graph, we find that a wave emerges that pulses progressively closer to the value of phi and recedes ever deeper into the fabric of reality, converging on 0.61803(. . .).

Conversely, phi also emerges when each number of the series is then divided by the one *before* it: 1/0, 1/1, 2/1, 3/2, 5/3, 8/5, 13/8, and so on. Intriguingly (at any rate, to me), here the first ratio of 1/0 is infinite, while the second and all subsequent terms ground the series in a finite form that this time ultimately converges on the alternative form of phi,

1.61803. . ., although the convergence progresses at a very slightly different rate to its inverse.

A beautiful example of the harmonic nature of phi was found in 2015 when astronomer William Ditto and his team at the University of Hawaii analyzed variable stars, whose brightness oscillates in a harmonic way. They discovered that sometimes not only were the two primary frequencies of the stellar "song" in phi ratio but also that for such "golden" stars the entirety of their harmonics displayed a fractal pattern.[9]

Phi's harmonic nature becomes even clearer when the progressive numbers of the Fibonacci series are expressed to form a special type of logarithmic spiral. These spirals are seen throughout nature, from nautilus shells and whirlpools, to the growth of plants and the embryonic development of animals, to the awesome scale of spiral galaxies.

Its presence pervades evolutionary and biological systems, and we ourselves embody relationships that approximate to phi in the ways the joints of our fingers and toes relate to our hands and feet and they in turn to our arms and legs and continue on to our whole body. Even our DNA is in phi ratio, with the width and length of a full cycle of its double-spiral molecule measuring twenty-one and thirty-four angstroms, respectively—both Fibonacci numbers.

The cosmic nature of phi and its harmonic presence isn't merely a feature of the Fibonacci series of integers. *All* series (such as the Lucas series which begins 2, 1, 3, 4, and so on) whose progressive integers, like Fibonacci's, are added to the one before and where each is then divided by its predecessor converge to phi—regardless of the first two numbers.

A number of geometers feel this is revealing something profound yet elusive about the deeper structure of reality. In recent years the mathematics of Fibonacci and Lucas functions has been combined with the so-called hyperbolic geometry that describes flat space-time. Named after Minkowski, who not only pioneered its understanding, as we've seen, but was also Einstein's teacher, such geometry (which enables events and their causality in space-time to be tracked) is the one in which his special theory of relativity is most often formulated.

While the real-world significance of Fibonacci and Lucas hyperbolic functions (I know, another snappy name) is an open question, I suspect it will be found to have important connections with the beginning and ongoing expansion of our Universe.

Watch this space-time . . .

INFLUENCES: SHAPE, SIZE, AND STICKINESS

The two key characteristics of manifestly very different phenomena that have probably surprised researchers the most are, first, the very widespread commonality of the patterns and relationships underlying their varied appearances, and, second, how apparently complicated real-world structures and systems can be effectively computer-modeled using simple rules.

Both these unexpected results began to be uncovered in the late 1960s. By 1970, an insight of what might be going on was reported by physicist Leo Kadanoff through studying the behavior of critical points at the edges of order and chaos for phase transitions, such as when water vapor either cools down or is put under increased pressure when it condenses to form liquid water.[10]

In the late 1980s more groundbreaking work was carried out by physicists Per Bak and his colleagues Chao Tang and Kurt Wiesenfeld at the Brookhaven National Laboratory on Long Island, New York, who decided to play with piles of sand. Increasing the height of the pile by one grain of sand at a time, at some point the addition of a single extra grain (to mix metaphors: "The straw that broke the camel's back") caused the pile to collapse. At this critical point, they measured how the collapse could be small, with only a few grains falling down the slope, or result in an avalanche. Rebuilding the pile again and again, they continued to plot other small and large collapses.[11]

What Kadanoff, Bak's team, and further researchers showed, using other materials and studying both real systems and by running computer simulations, was that the behavior of the ongoing dynamics between order and disorder in such systems is in principle exactly the same, completely regardless of their apparent diversity. It's due *only* to how easy or difficult it is for an informational ordering or disordering influence at one point in the system to bring order or disorder to another nearby.

And it's *only* the basic shape and physical size of the elements of the system that matter and that affect the type and scale of this influence, nothing else. In other words, shape, size, and their consequential stickiness are the sole factors that need to be considered to fundamentally understand and model the system.

What this universality also tells us is that any circumstances that embody the same simple criteria will behave in a similar fashion, regardless of their apparent dissimilarity. Such varied systems belong to a so-called universal class, and when the characteristics of one member of the class are understood, all other members of the same class can be treated in the same way.

And because the state of the system only appears to depend on the basic factors of shape, size, and stickiness, to effectively model its behavior one can simply exclude all other details, however complicated, that make up its appearance.

COMPLEXITY

The scale invariance and harmonic order of fractal patterns, power laws, and the spread of systemic influences are also aspects of the features of so-called complex systems, such as those we've already looked at including earthquakes and landscape formation. Here we'll bring these and other features together to describe the general behavior of such systems and how they can be understood in informational terms.

And in the next two chapters we'll explore them more specifically, first in terms of evolutionary biology and ecology and then in the socio-economic systems that we experience in our everyday lives.

To begin with, their complexity generally develops step-by-step and through nonlinear processes, where a tiny event can trigger either a small dislocation or a major upheaval or vice versa, and which takes them far beyond equilibrium to a critical state, where they exist at the edge between order and chaos.

The spread of such nonlinear influences of order and disorder is crucial to their behavior. Not only, however, do such influences take time to effect, in many instances they're also not one-way processes. Instead they can involve both positive and negative feedback loops that either augment or damp down the initiating influence.

The study of complexity emerged from earlier research into chaotic systems, whose outcomes are the result of relatively few nonlinear interactions and whose conduct is less governed by the past, until they pass through a threshold point (such as a specific temperature or a level of turbulence) when they tip into chaos. Complex systems, though, embody numerous interactions, and for them history definitely matters. All events in their past accumulate to form their present, and all current events irreversibly contribute to their future.

In contrast to chaos, studying complexity is really about trying to understand how simple behavioral patterns can somehow self-organize to lead to an enormous number of often complicated and dynamic relationships that despite being at the critical edge of chaos are, nonetheless, able to exist on a sustainable basis. They can do this as they're informationally underpinned by a fractal attractor within whose parameters a robust variability of factors can continue to maintain the system—albeit within limits.

One such complex system is the weather, whose underlying attractor is named after meteorologist Edward Lorenz and whose mapping in the complex plane reveals the beautifully symbolic shape of a butterfly. While Lorenz's first computer model depicted chaotic behavior, in his recognizing the longer-term stability of weather patterns he decided to look again for complexity. In doing so, he actually simplified his math, ending up with three short nonlinear equations from which a complex dynamic, yet inherently robust, system emerges.

The specific state of the system at any one time can be *anywhere* within the limits of the attractor, and the motion of the system never exactly repeats itself. This led him to coin the term "butterfly effect" in his seminal 1972 paper titled "Predictability: Does the Flap of a Butterfly's Wings in Brazil Set Off a Tornado in Texas?"—recognizing that very small differences in initial conditions can produce very different outcomes.[12] This also means that despite the enormously more powerful processing power of computers since the 1970s the weather still can't be forecast for more than a few days ahead.

You might be asking that if weather can't be forecast, how can climate change be anticipated? The reason is that while *precise* short-term weather behaviors are unpredictable owing to the sensitivity of future outcomes to specific initial conditions, by statistically averaging these out, much more foreseeable longer-term trends can be discerned. There's an elephant in the room with such long-term expectations, however, and as we'll soon see it's incompatible with the survival of the butterfly.

Before we do though we need to ask a key question: *How* do complex systems evolve to enable the robust levels of sustainability they embody? Early insights were again provided by Bak, Tang, and Wiesenfeld in their work on sand piles in the late 1980s (and then follow-up investigations on piles of rice by other researchers—honestly). Continuing to add grains to the accumulated piles, collapses both minor and major persisted. A stage, however, came when as more grains were added, on average the same number of grains toppled off the pile and so the amount of sand in the pile remained constant.

Before their groundbreaking work, other researchers had worked with systems whose properties required them to experimentally "tune"—that is, manually alter—the coefficients of the system, to reach critical points, such as the reduction in temperature or increase in pressure needed for water vapor to transition to liquid water. What astonished Bak and his colleagues was that the sand piles reached this point without any apparent external regulation, and so they described the system as having developed a property of "self-organized" criticality. In their breakthrough paper of 1987, they also stressed that the emergence of such criticality wasn't dependent on the exact details of the system and allowed flexibility and variability in reaching its critical state.

While providing significant insights into such complex behavior and the characteristics of the processes that bring it about, they and indeed many other researchers who continue striving to understand such systems are still trying to determine what it is that's self-organizing about them, how they do it, and what's happening in them that's different from other, noncomplex, systems?

To get closer to answering these fundamental questions, we need to understand how energy and entropy, and, most importantly, information, flow through these structures. Pioneering work by physicist Edwin Jaynes in the 1960s, physical chemist Ilya Prigogine in the 1970s, and many other scientists since have contributed to the following general insights that nonetheless are still being debated as active research continues.

Key attributes of such complexity include, first, that such systems are always dynamic and open to slow and steady flows of energy through them and, while far from equilibrium, are able to sustain energetic stability and so can exist in a semi-steady state.

Second, causality prevails throughout such systems, and history matters, as past events are inherent drivers of their future behavior. In other words, such systems are reproducible and can evolve.

Third, they embody long-range interactions that optimize their interconnectivity.

Fourth, the selection principle that seems to apply to their developmental states is that—subject to any externally imposed constraints—the flow of informational entropy is maximized. The operation of this principle also tends to most efficiently organize their innate harmonic order and therefore their levels of systemic coherence and resonance. Thereafter, as semi-stable organizations, informational flow gradually decreases to a minimum, ultimately resulting in an unchanging state, stagnation, or death.

Finally, such systems optimize the distribution of information expressed as energy, with their distributed intelligence then enabling them to be both robust and capable of repairing extensive damage to themselves.

BREAKDOWN OR BREAKTHROUGH

Even when such systems are robust, however, circumstances can arise where their ability to respond and adapt is severely compromised, and they're forced to undergo what are called critical transitions. Even though they are sustainable, nonetheless self-organized complex systems exist in critical states—on the edge between order and the turbulence of chaos. Yet their intrinsic attributes enable them to remain stable within the limits of their fractal attractors, often for considerable periods of time.

As more and more evidence mounts, it's becoming ever clearer, though, that many, and indeed perhaps all, such systems have critical thresholds, one or more tipping points beyond which the system inevitably and often suddenly transforms from one state into another.

With such a system's innate nonlinearity of causes and effects, trying to predict such critical transitions is extremely difficult, the more so because its organization may show little significant change before the tipping point is reached, but after which, like Tom chasing Jerry off a cliff, there's no going back.

However, given that such complex structures include everything from financial markets and social structures to ecosystems and our global climate, identifying stresses and early-warning signals prior to tipping-point breakdowns is extremely important to our collective well-being. For this, to take an urgent example, is where the elephant of possibly unstoppable and catastrophic climate change charges into the room, wreaking havoc and potentially crushing the butterfly.

In computer models and experiments with the simplest of complex systems, there do, however, appear to be a few features of behavior that characterize the approach of systemic breakdown. The first is a phenomenon known as critical slowing down (or CSD) during which the system recovers progressively slowly from small perturbations, rather like someone with a compromised immune system finding it harder and harder to recover from minor health issues.

As for a person's health, such CSD behavior tends to begin quite far away from what eventually is the tipping point after which a critical shift in the system is inevitable. Its continuation, though, leads to changes that constitute a second type of process that is called autocorrelation. This relates to the presence of repeating patterns over time of fluctuations within the system's behavior, and when a tipping point is approached the level of pattern repetition diverges from the norm.

The third sign, which arises from instability at the limits of what the existing fractal attractor of the system can handle, is that it becomes more variable, with progressively skewed changeability in its behavior.

Systems that are sufficiently destabilized may also exhibit what's known as "flickering," where the system wobbles between alternative states before undergoing a sudden transition to one or the other of them.

Given their importance, a great deal of research is being focused on identifying and finding ways to measure these early-warning signals, but there are enormous difficulties given their levels of subtlety and uncertainty and the challenges of studying real-world examples.

Political dysfunction, though, concerning public debate and policy regarding climate change imperils not only effective global action in responding to the currently perceived longer-term threats, but leaves us helpless in the face of such potential catastrophic breakdowns.

EMERGENCE

Together, what all these features of complexity demonstrate is that from the simplest (but no simpler) informational instructions encoding universal principles of geometry and scale and specifically timed and quantified flows of energized information, such complicated structures can arise. So it shouldn't come as a surprise that the most prevalent self-organized complexity is found in biological organisms and ecosystems and our collective social and economic systems, which is where we'll turn in the coming chapters.

For it's these most evolved of structures we yet know of in our Universe that most clearly embody the evolutionary process known as emergence, where so-called holarchic relationships are embodied in progressively complex organizations. Properties and principles describing one level of such a complex system don't necessarily explain another, higher level, despite how innately connected the two may be. The properties of atoms, for example, don't predict the emergent structure and behavior of molecules, which themselves don't enable the prediction of the emergent conduct of cells, whose activities don't encompass those of organs or the entirety of a biological entity. Understanding such emergence, where the whole is greater than the sum of its parts, and the evolved state is more organized than the totality of its components, is an exciting, continuing, and indeed contentious quest, as we'll now see.

9
In-formed Design *for* Evolution

Specification that when manifested achieves a specific goal: in this case, evolution . . .

> *DNA is like a computer program but far, far more advanced than any software ever created.*
>
> Bill Gates, philanthropist and cofounder of Microsoft

Nowhere else in our as yet discovered Universe reveals the presence of underlying in-formation as powerfully as the evolution of biological life-forms. The principles governing such emergence of increasing complexity continue, though, to engender polarized perspectives for their cause.

Religious proponents speak of intelligent design, whereas scientists talk of self-organization. By considering evolution, as for everything we call physical reality, as underlain by, pervaded by, and indeed made up of in-formation, we'll discover that such manifest self-organization and emergence can be viewed as having a deeper fundament.

So, as we'll see, instead of either intelligent design or self-organization, we can reconcile the two opposing outlooks in an expanded perception of an in-formed and in-formational design *for* evolution.

IN-FORMED EMERGENCE

For its entire 13.8 billion year lifetime to date, our Universe has been undergoing a process of evolution of structures and entities with progressively greater levels of complexity and whose antecedents don't themselves possess such properties.

Over the past 4 billion years or so, our Solar System and planetary home too have nurtured the emergence and evolution of ever more complex and self-aware biological organisms and organizations.

In the mid-nineteenth century Charles Darwin and Alfred Wallace both independently came up with the idea of natural selection as being the basis for such biological development. However, while Darwin's seminal *On the Origin of Species* introduced the idea of optimizing the fitness between an organism and its surrounding environment as defining its evolutionary progress and showing ample evidence for it *within* a species, the one thing he wasn't able to explain was the actual origin of a separate species—an emergent phenomenon. As yet, no one else has either.

We've seen how complexity can arise and indeed informationally self-organize from simple principles of combining building blocks and interconnections, and many scientists don't see any deeper mystery in how such processes can jump over the apparent gaps that characterize the ubiquity of emergence. While at each successive level, not only additional complexity but also new properties, phenomena, and principles of behavior are embodied, their perspective is that there's nothing here to see folks, so move along. I disagree with them.

One of the main aspects of emergence is the apparently spontaneous higher-level ordering of a system where it breaks through to not only a more organized holistic and holarchic entity but also one that often follows different rules and always embodies significantly more complex behavior, correlations, and coherence.

Until now there's been little understanding of how this occurs; however, recent studies of information and entropy are beginning to offer important clues, revealing greater levels of underlying order, increased through-flows of information, and informationally guided entropic processes during these transitional shifts than have been hitherto appreciated.

A considerable insight came in 2003 when mathematical physicists James Crutchfield and David Feldman asked a seemingly simple question: How does an observer come to know in what *internal* state a process is?[1]

By informationally modeling the relationship between an observer and the process observed and describing it in terms of a communications or measurement channel, they uncovered a trinity of insights that are vital for the ongoing quest to understand emergence.

Their first key finding was that viewing seeming uncertainty actually hides an ignorance of deeper levels of order: literally, where complexity goes unseen, it shows up as apparent randomness. The second was that memory is embedded and stored in such hidden order. And third, as they further reported with colleague Carl McTague in 2008, they realized that as the tipping point of emergence is approached, there's increasing and converging levels of so-called excess entropy, indicating the presence of additional and transient information at these junctures.[2] Such short-lived and specific information, it would appear, guides the process of transition and involves communication and sharing of data between different parts of the system and between the system and its surroundings.

An earlier indication of this had come in a 2002 paper by philosopher of science Brian Skyrms, who was studying the phenomenon of so-called cheap talk in the wider context of game theory, where it pertains to pre-play signals that don't appear to have any subsequent impact on the game itself, as they're the up-front cheap talk that's effectively ignored.

When he applied such pre-play signals in an evolutionary setting though, Skyrms found something very different and concluded that such cheap talk definitely does matter.[3] By carrying transient information, such formative signals cause large changes in the relative sizes of the underlying basins of fractal attractors, and so predispose a particular outcome. As a consequence, evolutionary games with such signals have and develop enormously different behavior and dynamics from the same games without them.

Given the informational sharing between the pre-emergent entity and its milieu, this offers a means by which those evolutionary directions that increase their fitness to their surroundings are favored. It also means, as indeed ever-increasing levels of evidence is showing, that an entity and its environs are essentially co-creative and evolutionary partners.

A paper in 2013 by systems theorists John Johnson, Andreas Tolk, and Andres Sousa-Poza further supports the crucial role of informational entropy in underlying and steering the process of emergence.[4] Researching systems of systems (SoS) whose different subsystems interact in complex ways and are underpinned by a series of multi-fractal attractors, they discovered that when an SoS approaches a critical point, the variety of its micro and macro states increases and results in emergence. Again, this argues for the presence of specific and transient information as harbingers of such leaps in complexity and also points to intrinsic co-creativity on wider levels of involvement than have so far been generally considered.

While understanding the underlying and innate presence of information throughout our Universe and its dynamic capacity to instruct the evolution of increasing complexity and emergence is still in its infancy, the developing paradigm of the cosmic hologram is offering us deeper insights into the nature of life and consciousness than ever before. Let's now turn to see how this increasing perception is rewriting the story of us.

THE FIRST TEN BILLION YEARS OR SO

We've already explored the informational instructions that underlie the appearance of energy-matter and space-time and how their incredible precision and correlations have enabled stars and galaxies to form. But such exquisite balance and exactitude didn't stop there: the active intelligence of our Universe has, instead, continued to ramp up, as we've seen, with the self-organization of more complex structures and the fractal patterns that underlie and in-form their manifestation. For such informationally guided evolution though to lead to biological entities, specific and additional building materials have been and are required—in the right amounts, in the right places, at the right times, and under the right circumstances.

So we'll now continue the story of our Universe, by focusing on this perspective.

Our bodies are composed primarily of four elements: hydrogen, carbon, nitrogen, and oxygen. While a number of others, including some in only trace amounts, perform vital functions, it's these main four that make up around 96 percent of the total.

The lightest of the four, hydrogen, was formed within the first few minutes after the beginning of the big breath (together with a far lesser amount of helium and tiny amounts of lithium and beryllium). The other three, however, and indeed all such heavier elements, are produced during what's generally known as nucleosynthesis, which takes place in the interiors of stars as they age, during the cataclysms of supernovae explosions, or in extreme high-energy impacts between cosmic rays.

Stars are born, live, age, and die as we do. Within their hot and pressured interiors, hydrogen nuclei, which make up most of their initial mass, progressively fuse to form helium thereby releasing energy and enabling them to shine. Eventually as the star ages, its hydrogen fuel becomes depleted, other processes take over, and heavier elements including carbon, nitrogen, and oxygen are synthesized and build up in the outer layers of its body. Often at the end of the lives of small to medium-size stars, such as our own Sun, these outer layers are then ejected to enrich the interstellar medium.

For larger stars, the enormous forces involved continue the nucleosynthesis process to form still heavier elements. Energetically building up to a point of eruption, most of their mass is then blasted far out into space in the immensity of a supernova explosion.

The forces and dynamics that shape galaxies such as their dark matter substructure, their spin—especially with regard to spiral galaxies—and their electromagnetic fields result in the clustering of such star-seeded material into interstellar gas and dust clouds. It's within these that the gestations of the next generations of stars and their planetary systems are nurtured. When described in this way, it seems relatively straightforward, but I hope a few additional considerations will reveal just how wonderfully special this further deep ordering of our Universe is.

To begin, let's consider why the clustering and seeding of interstellar clouds occur as they do. As our Universe expanded and cooled, the first galaxies began to form, only a few hundred million years after its birth, from the gravitational contraction of primordial gas. As they did so, they started to spin, owing to small density variations leading to differences in the rates of infall at different points. These differences translated into rotational energy and caused the gas to then cluster along spiral arms that we're familiar with from our own Milky Way galaxy. Such organization of visible energy-matter formed the foam on the surface of the hidden substrate of dark matter, itself also affected by the slight differences in gravitational attraction.

So the scene was set for further gathering of matter in the forms of the first generation of stars. With so much available hydrogen and in close proximity to each other, these first stellar cohorts were massive. They lived hard and died young with their demise, perhaps in coordinated bursts of supernovae explosions, abundantly seeding the interstellar medium. With the next one or two generations of stars continuing to provide an ever-richer elemental heritage for their successors, the interstellar clouds of gas and dust grew and prospered to become incredible birthing places not only for new stars but also for their future planetary systems. One of the most iconic images captured by the Hubble telescope in fact shows such a stellar incubator. In 1995 it took an exquisite photo of globules of embryonic stars, dubbed EGGS (evaporating gas globules) and appropriately discovered in the "nest" of the Eagle Nebula, a nearby star nursery some 6,500 light-years away from us.[5]

Two further factors combine to play their roles in ensuring the optimum conditions for such midwifery. On average, in our galaxy, three supernovae occur every century, and their explosions send out enormous shock waves whose specific frequency and power is key. Not only do interstellar clouds seem to be most efficiently created at their intersections, but subsequent waves sweep up the material and trigger the formation of new stars as well. Once again, we find that their combined rate and force are perfect: if more frequent and/or powerful, the interstellar clouds would be too turbulent for birthing future stars; if less frequent and/or powerful, the cloud material would disperse before stars could form.

The second critical aspect, as discovered in 2009 by astronomer Hua-bai Li, then at the Harvard-Smithsonian Center for Astrophysics, and his team is the presence of strong and coherent magnetic fields within and between such gas clouds.[6] Aligned across both small and large scales, these act as an inhibitor to further gravitational attraction and influence the sizes of the stars and associated planetary systems and the timings of their birth. In doing so the longevity of stellar lifetimes is such as to enable the evolution of biological complexity within planetary systems, and the overall active lifetimes of galaxies are extended beyond what they otherwise would be.

Having discussed the marvelously balanced conditions for the birthing of solar systems, the components of the gestation clouds are no less perfectly composed to enable the eventual emergence of biological life. Over the past few years, astronomers have increasingly discovered a broad range of organic molecules in interstellar clouds. In 2014 a group whose lead author was Arnaud Belloche from the Max Planck Institute for Radio Astronomy reported the most complex so far: isopropyl cyanide.[7] Significantly, its branched carbon structure is also exhibited by amino acids and the bases of proteins, and its detection implies the widespread presence of such prebiotic building blocks. It was found in the giant gas cloud near the center of our galaxy that has also been shown to include vinyl alcohol and ethyl formate molecules, chemicals that give raspberries their flavor and rum its smell, suggesting the presence of a boozy dessert in a very unexpected place.

The other essential requirement for, and component of, biological life is water, and astronomers have also discovered vast amounts of ice in interstellar clouds that form at temperatures only ten degrees above absolute zero. In 2014 a team including Tim Harries at the University of Exeter in the UK analyzed the makeup of Earth-based water and compared it to that in such clouds, reporting that up to half of the water on our planet was, amazingly, formed before the birth of our Sun.[8] We're not only formed of ancient stardust, but we also drink it!

So, within the clouds, the components of life are present. What, though, then energizes their further self-organization to form the fundamentals of biological entities, the carbon-based organic molecules that have been observed in these clouds? It's the presence of light and specifically of ultraviolet frequencies. When low-level UV light irradiates the ice-chemical grains that pervade the clouds, it provides the ideal and vital energetic trigger for the construction of carbon-carbon bonds as a pathway to such emergence, as first reported in 2002 by astrobiologists Tania Mahajan, Jamie Elsila, David Deamer, and Richard Zare.[9]

By the birthing of third- or fourth-generation stars, such as our Sun, everything was then in place to enable the formation of planetary solar systems that not only include light gas giants such as our own Jupiter and Saturn but also heavier rocky planets such as our Earth. Offering the possibilities of being perfect havens for the emergence and evolution of biological life-forms, the scene was then set for the next steps of the journey to us.

Nearly five billion years ago, embedded in a spiral arm of our galaxy about two-thirds of its full radius out from the supermassive black hole at its center, a small area embedded within a larger cloud of gas and dust began to gravitationally collapse. Likely triggered by nearby supernovae explosions, within the next couple of million years, perhaps less, the central matter had coalesced and heated up to form our Sun, surrounded by a rotating disc of protoplanetary material within which recent evidence points to further development of complexity having continued.

In 2011 a report based on studies of meteorites by NASA suggested that RNA, ribonucleic acid and the forerunner of DNA, may have formed extra-terrestrially.[10] And in 2012, astronomers detected glycolaldehyde, a sugar molecule required to form RNA, in a proto-stellar system.[11]

Over the next few million years, further gravitational accretion within the spinning disc aggregated tiny planetesimals, then larger planetoids, and finally planets and moons. Heavier rocky planets formed closer in, and lighter gas giants such as Jupiter and Saturn farther out. Indeed Saturn is so light that if there were an ocean large enough, it would actually float.

With planetary science progressing in leaps and bounds over the past thirty years or so, many previous ideas of how our Solar System came about and evolved have been thrown out; new ideas have been developed and incorporated into an increasingly consistent picture, and indeed some earlier discredited ideas are being rehabilitated. Importantly, however, it's becoming clear that, as planetary astronomer Renu Malhotra has noted, "Solar system dynamics is a tale of orbital resonance phenomena . . . [which] can be the source of both instability and long-term stability."[12]

As long ago as the eighteenth century, astronomers Johann Titius and Johann Bode, among other near contemporaries, realized that the relative orbital sizes of the planets then known, from Mercury out to Saturn, were indeed in close resonance. Later known as the Titius-Bode law, the rule was used to correctly predict the existence and orbital radii of both the asteroid belt and Uranus. Later discoveries of Neptune and Pluto, though, whose orbits didn't follow the rule, caused its premise to fall out of favor. Recently, however, it's been determined that the early planetary orbits of our Solar System migrated. As far out as Neptune, owing to various factors, the planets moved either closer to, or farther away from, our Sun to eventually settle into the resonant and long-term stable orbits that we see today.

Not only are the insights of Titius and Bode being re-appreciated, but other and subtly different resonances are also being discovered: for instance, where planets are in so-called resonant-angle libration with each other, which occurs when their angular separation, as they move around their orbits, oscillates between certain boundaries in resonant cycles. A few examples of such resonant pairings in the outer planets of our Solar System are that the orbits of Jupiter and Saturn are within 1 percent precision of 5 to 2, Saturn and Uranus within 5 percent of 3 to 1, and Uranus and Neptune within 2 percent of 2 to 1 exact resonances.

The orbits of Neptune and Pluto, whose specific contravention of the Titius-Bode law led to its disfavor, have also been found to resonate, primarily of a 2-to-3 ratio in their orbital periods, which together with other influences has evolved to predominate their paths around our Sun.

Further insights, and specifically about the key roles of Jupiter and Saturn, have revealed too how grateful we should be that these two giants are members of our solar family. It's been known for some time that Jupiter's great gravitational pull helps to sweep the inner Solar System clear and so fortunately keeps our own planet relatively (but not completely) free from dangerous space debris. However, in March 2015, Jupiter's role was shown to be even more significant in its early years.

A paper by an astronomical team whose coauthor was Gregor Laughlin at the University of California–Santa Cruz also helps explain why our own Solar System is quite different from the other multiple planetary systems so far discovered.[13] Most of them, in contrast to ours, appear to contain very large so-called super-Earths orbiting very close to their stars and so would probably be non-habitable. In addition, within our own system, recent evidence shows that Earth and other inner rocky planets formed later than the outer gas giants, including Jupiter.

According to Laughlin and his coauthor Konstantin Batygin, Jupiter swept through the early Solar System. In a tumultuous game of cosmic pinball it actually destroyed a first-generation of inner rocky planets that would have gone on to become super-Earths. Having been as close or even closer to our Sun than the current orbit of Mercury, their heavier enriched remnants then fell into its fiery embrace.

Building on an earlier proposal of 2011 by another team and called the grand tack hypothesis, their model also shows that after its spree, it was Saturn's gravitational pull and perhaps a resonant connection between them that would have pulled Jupiter back out to its current orbit, thereby allowing a second generation of inner rocky planets to form. From the depleted material that remained came Mercury, Venus, Earth, and Mars, which are less massive and with much thinner atmospheres than would otherwise have been the case.

Saturn is important in a further way as is strongly suggested by a modeling of planetary orbits carried out by Elke Pilat-Lohinger of the University of Vienna in 2014.[14] Earth's orbit is stable and nearly circular, with our distance from the Sun only varying by around 2 percent, thereby ensuring that we remain in what's known as the habitable, or Goldilocks, zone, not too hot or too cold, and where liquid water vital for biological life is present. As she found, moving Saturn's orbit 10 percent closer to the Sun or tilting it by about 20 percent would significantly elongate and disrupt our own, causing a tug that would pull us way beyond the habitable zone for part of each year and so making the evolution of life on Earth much more challenging, if not untenable.

On the first of August 2008, I climbed one of the peaks of Mount Huashan near Xian in China to witness a total solar eclipse. Across from me was another of the five peaks of this sacred Lotus Mountain above which the Moon's shadow was slowly extending across the disc of the Sun. As both continued on their course, at the first moment of totality, I experienced one of the greatest of natural phenomena. This time, though, it was even more spectacular as the combined solar-lunar disc was also, from my lookout, perfectly and exactly balanced on the peak opposite; the cosmic trinity of Sun, Moon, and Earth was revealed in the most exquisite sight I've ever seen.

Our Sun is four hundred times larger than our Moon; it's also precisely four hundred times farther away. It's this extraordinary correspondence, unique in our Solar System, that enables total eclipses to occur.

We've seen briefly how our distance from the Sun, occupying the Goldilocks zone, allows biological life on Earth to prosper. However, the presence of our Moon, Luna, in its size and proximity to us, also plays a major role. Our Moon is far larger in proportion to Earth than any other moon to its planet in our Solar System other than the astronomically arbitrated "dwarf" planet Pluto and its largest moon, Charon. The tides caused by its gravitational pull help dissipate heat and nutrients in our oceans and seas, optimizing energy flows, nurturing the widest and most abundant food chains, and likely speeding up speciation and diversity. Its mass also helps to steady our axis of rotation, thereby reducing the scale of inherent climate change. Its presence has also slowed Earth's rotation down from what it originally was, perhaps as fast as less than three hours to a day, and so again enable more stable conditions to exist. Given the commonality of terrestrial and lunar rocks, with their identically measured oxygen isotopes, current debate generally considers our Earth-Moon system as having been the result of a cataclysmic impact in the very early days of our planet's story. In September 2016 a study published in the journal *Nature Geoscience* by Johns Hopkins University researchers provided further strong evidence for such a scenario by showing how a stratified layer of iron and other elements deep within the Earth was very likely to have been formed by turbulence caused by such an impact.[15]

In addition, modeling collision scenarios by an external impactor hitting the early Earth all point both to the Moon consequently forming much closer and our planet to have been spinning much faster than now.

The combining of scientific disciplines is also enabling an ever-greater appreciation of how the Sun, Moon, and Earth interact, especially through the finely tuned interplay of electromagnetic forces, to protect and energize life on Earth. Our Sun's heliosphere, the huge bubble of magnetic force it emits, encompasses our entire Solar System. As we orbit the center of our galaxy, its electromagnetic halo protects us from lethal levels of cosmic rays and the pressure, turbulence, and roving debris of the interstellar medium. However, the solar wind, the electrically charged plasma that streams out of the Sun to maintain the heliosphere, carries its own dangerous energies. We in turn are protected by our Earth's own bubble of magnetism, the magnetosphere, this time perpetuated by the inner geo-dynamo generated by our planet's iron-rich core.

While the Moon's magnetic field is now very weak, around 3.7 billion years ago, at the time when biological life was starting to establish itself on Earth, it was as strong as ours is now. As yet, little is understood as to how the Moon initially produced such a large field or how and why it then so substantially decreased. However, the combination of these three factors during that vital epoch, I think, offers an intriguing possibility.

In the presence of water and organic molecules, already abundant on the surface of the early Earth, could powerful electromagnetic interactions have caused massive geomagnetic storms in our planet's then volatile atmosphere? If so, these might have acted as a trigger not only for the further self-organization of prebiotic organic molecules but also for the critical emergence of the first biological molecules and, most likely, RNA.

Throughout the entire history of Earth, these and numerous other interactions of EM fields between our planet, our Sun, and our Moon, and the fundamental information they carry and communicate, have and are crucial to all biological life.

It's to this emergence and evolution that we'll now turn.

THE EARLY DAYS OF EARTH

The age of our Earth is currently reckoned to be just more than 4.5 billion years. Initially molten, it cooled to form a solid crust while then being pummeled with frequent collisions by asteroids and comets. Given the abundant presence of water and prebiotic organic molecules in the gas clouds from which our Solar System was born, and even the potential formation of RNA, such impacts may well have delivered vital ingredients for organic life. The growing evidence for such a possibility, known as *panspermia* and previously discounted, is now being considered as an increasingly realistic scenario.

In any event, by 3.8 billion years ago, or even earlier, our planet was cool enough and stable enough to be ready for such life to emerge—whether formed indigenously or by its earliest immigrants.

While solving the mystery of its specific emergence remains elusive, in general it's becoming ever-more clear that informational storage, pathways, and flows are fundamental. As we've seen, additional, explicit, and transient information is characteristically present just before emergence itself and involves an informational dialogue among the pre-emergent entity, its holarchic subsystems, and its larger milieu.

For some time, biologists have generally viewed RNA as being vital in such emergence from prebiotic organic molecules. Like its more efficient and stable successor DNA, RNA's ability to store and process information, be self-replicating, and enable the coding, regulation, and expression of genes and the formation of proteins is key. The challenge, though, is in understanding how such a complex entity as RNA could have assembled itself from its constituent components in the conditions of the early Earth.

In June 2015 two potentially groundbreaking studies on the formation of RNA, one led by Charles Carter and the other by Richard Wolfenden, both of the University of North Carolina, and published in the *Proceedings of the National Academy of Sciences* offered glimpses of how such emergence might have arisen.[16]

RNA is made up of a collection of four subunits composed of organic molecules called nucleotides, and the question Carter and Wolfenden and their teams were seeking to answer is how they came together in the chemical soup of the primordial Earth.

Whereas, in May 2015, chemist John Sutherland and a team at the University of Cambridge in the UK showed that a cyanide-based chemistry could make two of the four subunits and many amino acids,[17] there was still no mechanism for then assembling the nucleotides into RNA, for amino acids to form, or indeed for RNA to guide (or code for) the formation of proteins—another vital building block of biological life.

Carter's study tried to answer how RNA makes proteins by looking at how a molecule known as transfer RNA, or tRNA, reacts with different amino acids. We saw awhile ago that the in-formation embodied by shape, size, and stickiness are the sole factors that affect how different elements in a system influence each other. Guess what? His team found that one end of the tRNA sorted amino acids by their shape and size, while the other end could connect with amino acids of a certain electrical polarity, affecting that end's stickiness. So the tRNA directed both which amino acids combine to form a protein and its final shape, essentially an intermediate step in the development of genetic coding.

In the other study, Wolfenden and his team investigated how amino acids distribute in water at the high temperatures existing on the primordial Earth. Again they discovered that it was their shape, size, and polarity (stickiness) that mattered. The research also concluded that high temperatures still resulted in the same rules for the geometry of protein shaping, known as folding; the information was being sustained and so was able to be replicated.

ELEC-TRICKERY

From the primordial light that suffused our early Universe, the electric and magnetic fields and electrically charged plasma that permeate space and help guide the formation of galaxies and stars, to its planetary role in helping sustain life on Earth and its being vital to our global technologies (called elec-trickery by my bemused grandma), electromagnetism is an ideal, essential, and universal informational tool of the cosmic hologram.

Its ability to store, process, and communicate maximal amounts of information is especially significant for processes that involve the complexity that biological emergence and evolution embody. Indeed the difference between a living biological entity and a dead one is essentially the cessation of electromagnetic activity in the body.

In the forms of electrical gradients, potentials, and charge differentials, EM fields, usually on small scales and often transient, guide numerous biological activities such as the firing of neurons in the brain, muscle flexing, and hormone secretions, which in turn drive many other processes. Most of the molecules in the bodies of organisms react weakly with extremely low-frequency EM signals, and on an even more general level, electrical flows energize intercellular and intracellular communication throughout the bodies of biological organisms.

Magnetic fields, whether pervasive or pulsed, are also endemic to biological processes; perhaps best known to date are the navigational abilities of migratory birds, animals, and insects that rely on Earth's magnetic field to safeguard their journeys, over often enormous distances.

While bio-electromagnetic research is still in its infancy, it is progressively revealing the involvement of low-level EM fields and energy flows in transferring information throughout the body. Although still controversial, there are indications that such fields may play an even more comprehensive role, as metaphysical traditions have maintained since ancient times, in determining the coherence and wholeness of healthy bodily templates.

ORGANIZATION

From accumulating insights into the deeper nature of the interconnections, self-organization, and development of biological organisms, broader and more general principles are beginning to be formulated. In terms of in-formational and entropic processes and involving energy transference they arise from deeper and nonphysical patterning, templates, and attractors and guide biological emergence and evolution.

As we've seen through our explorations so far, such foundations and their physical manifestations are essentially holographic and holarchic. So it should come as little surprise that biologists too are discerning the signature of the cosmic hologram. In this perspective, the informational templates of fractal attractors are being perceived to underlie biological forms too. Progressively more complex multi-fractal attractors form systems of systems in response to dynamic and co-creative evolutionary processes that intimately interconnect biological entities and their surroundings.

Studies of morphogenesis, of how the shapes of biological entities mature, is progressively recognizing the underlying in-formational processes that guide such development. This guidance is achieved through a range of signals and responses embedded in hormones, chemicals, EM fields, and electric charges and following a model coded by the DNA genome within every cell.

The cells of all biological organisms on Earth contain DNA. However, they encode different genomes, the complete set of genes that are specific to each species and that describe its form and function. In a human body, as in every other biological species, the same DNA of its entire genome is encoded in every cell. Yet beginning with undifferentiated stem cells, the underlying in-formational template, through the messaging of DNA, progressively ensures cellular specialization and guides the fetus to eventual maturity.

This informational underpinning and its bioelectrical processes are also being investigated to understand how to regenerate lost limbs, as some animals can do naturally, and even to potentially regrow diseased organs. Michael Levin and his colleagues at Tufts University have shown that Planaria flatworms aren't just able to regenerate their heads, but when they do so they remember information they knew before decapitation.[18]

Levin's team has discovered that electrical resting potentials are encoded across bodily tissues, and these determine what, when, and where tissues and organs are constructed. They've also uncovered that bioelectrical conversations at the cellular level and across a wide area, perhaps the entire body, mesh to direct the growth of its complex structures. As Levin has noted, "Focusing on a strategy that harnesses what the host organism already knows about how to build its organs is the way to go."

In the past fifteen years or so, the role of DNA itself and the genetic code have been radically overhauled and the preeminent stance of the "selfish gene" progressively relegated. The revised view is that not only are genes more servants than masters of bodily organization and evolutionary progress but also that organisms both within themselves and with each other are more cooperative than conflictual and that organization and evolution not only occur at the individual but also at larger group and collective levels.

This reappraisal really began to take shape back in 2001 with the outcome of the Human Genome Project, which aimed to map the entirety of our genetic makeup. Its results were completely unexpected. Instead of the hundred thousand or more genes assumed to be required to control, or, as biologists say, code for, the equivalent number of bodily proteins, it revealed our genome to comprise a much lower number. An estimate in 2015 shows our entire protein-coding genome to include fewer than twenty thousand genes, similar or even fewer than many simpler species, including worms and even some plants such as rice and onions. Ah well.

In addition, researchers have come to the realization that a massive 98.5 percent of our entire genome is noncoding for proteins. Previously considered a much smaller component and viewed as having no current biological use, it was previously denigrated as "junk" DNA. Begun in 2003, however, the Encyclopedia of DNA Elements', or ENCODE's, ongoing research aims to understand its true significance. Now more appropriately referred to as noncoding DNA, its full purpose is still being unraveled; in 2012 the consortium of ENCODE researchers identified that more than 80 percent is biochemically active, and much of it is involved in controlling the expression of the levels of coding DNA.

A further and basic premise of the gene-centered paradigm has been that nothing that happens to the organism during its life can affect its genes or be transmitted to future generations. Yet, this view too has now been discovered to be incorrect. Instead, there is increasing evidence for the role of so-called epigenetic influences where informational attributes of lifestyle and environmental factors can persist as stably inherited traits. These discoveries and ongoing explorations are leading to major rethinks, not only about the role of DNA but also the much broader implications of how informational flows, interactions, and memories organize and evolve biological entities.

We've seen how pre-biological organic molecules have been discovered in interstellar dust clouds, alchemically produced from basic elements and the presence of water and ultraviolet light. In 2001 astrobiologist

Louis Allamandola, leading a NASA team, re-created these conditions, as others have done since, in the laboratory.[19]

They discovered that the molecules spontaneously organized into tiny vesicles, organic sacs that are essentially membranes that separate and crucially protect the components inside from their outer environment. Rod-shaped, they have a further and critical feature. One end electromagnetically attracts water, and the other end repels it. Given the fundamental roles of both water and electromagnetism to all biological life, it's a highly effective way to self-organize and key to the initial survival and evolution of all life-forms.

Biological membranes, however, are not passive but actively mediate continuous two-way signaling processes between the environment and the organism. They do this via so-called receptor proteins embedded within them that recognize and respond to informational signals both chemical in nature, such as electrically charged ions, and energetic, for example, electromagnetic vibrations. So important are receptors to the construction of our bodies that it is estimated that some 40 percent of coding DNA is there to ensure that they are perfectly reproduced.

Other research teams are studying the vibrational responses of DNA to such information suggesting that its form as a wound double helix is ideal to then act as an antenna, receiving and sending signals to turn on or off the expression of genes. This too provides further insight that, rather than being the master key to our physical form, our DNA is the biological tool kit that mediates our thoughts and emotions and, as pioneering cell biologist Bruce Lipton has pointed out, our beliefs—whether they may be true or false.[20]

Combining a perspective of the rebalanced role of genes, the essential nature of membranes and the crucial presence of information for the evolution of biological life, Lipton takes the view that instead of cell nuclei (where most of cellular DNA is located) being the "brains" of the cell, it is their membranes that not only perceive the environment but also actively mediate the information entering and leaving. He and others have also noted that the optimal way of energetically organizing two-dimensional membranes in three-dimensional space is by embodying fractal geometry. Lipton describes the progressive embodiment of complexity as fractal evolution, where with the development of multicellular organisms, whole-body membranes surrounding the organism then act as holographic processors for the energetic and so informational templates of their overall forms.

Researching the multi-fractal foundations for biological morphology is also very much an active study. One fascinating conjecture, proposed by systems theorist Stuart Kauffman, is that the 256 specialist cells in the human body may be a system of systems (SoS) of an underlying and coherent network of the same number of multi-fractal attractors.[21] He's come to this conclusion after charting the amount of DNA in a specific cell against the number of different cell types for various organisms and has shown that they relate to each other as a power law, thus revealing their inherent fractal nature. This would be analogous to other SoS cases so far studied and would enable a more holistic and informational approach to the question of how intelligence may be distributed and organized around the body.

In 2010 neuroscientists Larry Swanson and Richard Thompson from the University of Southern California offered helpful insight when they injected molecular tracers at exact points within a small section of the brain tissue of a rat that previous research had shown to be associated with pleasure and reward. Instead of demonstrating the then neuroscientific consensus of to-and-fro signaling with a central processing hub, they instead saw a complex interconnected network linking regions not previously known to communicate with each other. Essentially the brain behaves more like the fractal Internet.[22]

Both the heart and gastrointestinal system have also been found, like the brain, to have their own neuronal cellular networks, connections of specialized cells that use electrical potentials across cell membranes to communicate signals. Neurons aren't specifically brain cells, and their much wider dispersion in key communities around the body argues for a deeper level of communication, whether conscious or autonomous. Such discoveries are offering new takes on having a sinking heart or a gut feeling and deeper insights into the distributed intelligence of our bodies.

We'll soon explore further what such distributed intelligence this is and where much other accumulating evidence may be pointing in shedding light on the nature of cognition and consciousness. Now, though, we'll consider in a little more detail how dynamic interactions between biological organisms and their environmental conditions drive evolution.

CO-CREATIVE EVOLUTION

When environments change quickly or even cataclysmically, the informational discourse with their biological expressions elicits novel responses. The underlying fractal attractors bifurcate into emergent new forms. Here is where the origin of species truly lies.

The fossil record, combined with other evidence of past climate changes, testifies to a history of the evolution of life on Earth being through a process of so-called punctuated equilibrium, which exactly reflects this developing informational perception. In periods of stability, such as our own Holocene era during which unusually quiescent environmental conditions have persisted for some twelve thousand years, only relatively minor evolutionary processes on intraspecies levels take place, just as Darwin found.

Large-scale climate cycles and longer-term deviations, such as ice ages and warmer periods between them, respond to the interplay of changes in Earth's movements. Known collectively as Milankovitch cycles after Serbian geophysicist Milutun Milanković, these changes are the result of Earth's orbital eccentricity and axial tilt and the precessional wobble around its axis.[23] Biological responses to these historically show the progressive die-off of preexisting species and the emergence of new, climate-adjusted variants, often with relatively close links to their antecedents. With ecosystems increasingly understood to exist in states of near criticality, such apparent instability actually confers a level of flexibility that is able to optimize such responses with the minimum amount of genome reconfiguration, thus enabling evolution to reap the maximum benefit with the least effort.

However, it's catastrophic changes that elicit a whole new scale and speed of evolutionary development—possibly the most extreme case of the adage that what doesn't kill you makes you stronger. Geologists have confirmed the occurrence of at least five near-extinction events in the past 540 million years during each of which more than 50 percent of extant animal species perished.

The event of 252 million years ago and known as the Great Dying was the largest, with more than 90 percent of all animal species killed. It had enormous evolutionary consequences, though, in the resultant spawning of whole new and more complex types of organisms.

The latest, the wipeout of around 66 million years ago, whose primary cause is generally considered to have been an asteroid impact in what is now the Straits of Yucatán in Mexico, saw the demise of the dinosaurs and the emergence of the mammals.

After each event, not only was there recovery but also rates of emergence and increased complexity were dramatically boosted as innovative signal-response communications within environment and ecosystems prevailed. More than 99 percent of all biological species that have ever lived on Earth are estimated to be extinct. The 1 percent now inhabiting our planet are themselves, however, under extreme threat.

Many researchers consider that another mass extinction, even faster than the Cretaceous-Paleogene boundary event that killed the dinosaurs, is happening now. Unlike past events with natural and often multiple causes, there is wide consensus of there being just one reason for today's devastation—us.

LIFE WILL FIND A WAY

From the first moment of the big breath, the foundational information of our Universe and its entropically increasing dynamism has enabled and nurtured the perfect circumstances for our Earth to evolve and embody increasing complexity, enormous diversity, and the progressive self-awareness of biological organisms. The story of our planetary history and the emergence of ourselves are extraordinary. Here I've only told a small part to illustrate some general principles of the cosmic hologram.

Over the past few years, astrobiologists have been simultaneously astonished and excited to discover organic molecules and water forming in the unexpected settings of interstellar gas and dust clouds. Here on Earth, biologists have been equally surprised to find viable life-forms, called "extremophiles," living and indeed thriving both inside the torrid conditions of active volcanoes as well as miles down in the dark, cold, and highly pressured depths of the oceans, organisms that are able to withstand enormous levels of high and low temperature, pressure, acidity, and radiation. While such conditions haven't enabled their further evolution, many of these fellow creatures may be some of our oldest ancestors.

We've seen, though, how being in the Goldilocks zone and sustaining more benevolent environments, Earth has also been able to nurture the evolution of more complex biological life for nearly 4 billion years. Recently, astrobiologists are also coming to realize that where planets or moons generate sufficient inner heat, enabling liquid water to form in a stable setting, life may be able to emerge even beyond such ostensibly habitable regions. In our own solar system, two part-rocky moons—Europa, a satellite of Jupiter, and Enceladus, a moon of Saturn—may both harbor life in warm subsurface oceans sheltering under frozen crusts.

Beyond our Solar System, by the end of 2016 more than three and a half thousand exoplanets were known to be orbiting other stars in our galaxy, with some six hundred hosting multiple planetary systems. One in five sun-like stars is also currently estimated to have an Earth-size planet occupying the Goldilocks zone. We've only just begun the search to find extraterrestrial biological life-forms. A generation ago, most scientists would have confidently stated that we're probably alone in the Universe. No longer is this so. It's becoming ever clearer that our Universe is most likely teeming with biological organisms. It seems that, if at all possible, life will find a way.

10
Holographic Behaviors

Human beings in both space and time act holographically . . .

> *For much of my life there was no place where the things I wanted to investigate were of interest to anyone.*
>
> Benoit Mandelbrot, mathematician and "father of fractals"

While we consider that our choices are of our personal volition, there's growing and sometimes astounding evidence that when aggregated, our collective behaviors also embody the all-pervasive signature of the cosmic hologram. Fractal patterns, self-similarity, scale invariance, harmonic resonances, and power-law expressions are continually being revealed as permeating our human-prescribed structures and organizations.

We'll now encounter some of the discoveries that confirm that, as is the case throughout the so-called natural world, the ubiquity of holographic in-formation imbues human-instigated phenomena as apparently disparate as the growth of cities and the interconnectivity of the Internet, as well as such contrasting occurrences as the incidence of conflicts and the everyday events of our social activities.

NO DIFFERENCE

So far, we've explored the nature of what we call physical reality as manifesting from a superphysical and all-pervasive foundation of information that is essentially co-created, directed, and infused with and by in-formation playing out holographically and dynamically and evolving throughout the space-time past, present, and future of our Universe.

We've discovered ever-growing evidence for the cosmic hologram and noted how its innate in-formation also underpins the progressive development of complexity, including the emergence and evolution of biological life-forms. Singular biological entities, such as each of us, think, feel, and experience in ways that are unique to ourselves and yet we share a commonality in terms of being human.

We've already seen an example of such collective holographic patterns when we touched on how Benoit Mandelbrot's research into the price movements of corporate stocks has shown them to embed a multifractal structure.

So, let's now turn to some more of the increasing number of examples that are demonstrating the holographic and holarchic realities about our group and collective behaviors and revealing that there's really no difference between ourselves and the rest of the world in terms of the manifest patterns of reality.

THE INTERNET

Nobody planned the global Internet. Back in the 1960s, its original structure followed ideas by engineer Paul Baran, who suggested, in contrast to the centralized approach then prevailing, a highly decentralized computer-based communications network. The distributed structure of such a network, where each computer node is connected to a number of others, incorporates a high level of redundancy, which results in a setup that is robust, flexible, and less vulnerable to attack or systemic breakdown.

In 1967, computer designer Wesley Clark took Baran's ideas to come up with an innovative database system for sharing information and which became the World Wide Web structure for websites.

More than twenty years later, Sir Tim Berners-Lee (affectionately often known as TimBL) added the third key attribute when he developed and freely offered the computer language of hypertext that enabled data to be uniquely tagged and hyperlinked throughout the system.

From these serendipitous but rudimentary beginnings the Net has continued to self-organize and develop with innately holographic characteristics. Let's look at a couple of the main ones.

Clustered in cities, scattered throughout rural regions, and spread unevenly across all time zones around the globe, early researchers presumed that the computer traffic of Net usage, at the individual or corporate behest of users, wouldn't display any such patterning.

However, in 1998 computer scientists Walter Willinger and Vern Paxson were the forerunners of a number of researchers investigating the statistics of such traffic across the World Wide Web over a period of time.[1] What theirs and subsequent studies have unexpectedly shown is that across a wide range of timescales, traffic is self-similar and possesses fractal properties.

While anyone who sets up a web page on the Net also has choices as to how many outgoing links they include, they can't control the number of links that their page attracts. So once again the anticipation was that there would be no patterning in the nature of such connections.

In 1999, pioneering network theorists Réka Albert, Hawoong Jeong, and Albert-Lászlo Barabási decided to test this assumption by counting the links of a Net-based database comprising some three hundred thousand documents and around one and a half million connective links.[2] What they found, despite the numerous unmanaged choices involved, was that its connectivity followed the holographically scale-invariant features of a power law. They also discovered the highly self-organizing and adaptive properties of the system and determined that, like ecosystems and many other complex phenomena, the Net develops in a sustainable but critical state.

The holographic nature of the Internet backbone, its principal data routes, was also proved in a 1999 study by three computer scientist brothers: Michalis, Petros, and Christos Faloutsos. They showed that such scale-free networking also applies to the physical structure that supports the Net: the node points of routers and their communication links with individual computer access points.[3]

According to research company eMarketer, the total number of Net users was nearly three and a half billion by the end of 2016. Just as for the evolution and emergence of biological complexity, the expansion of the Net has revealed emergent properties that can be modeled using precisely the same mathematical tools as those to study biological ecosystems.

Significant examples of its emergent behaviors are those of open-source software, where anyone can make changes. The growth and success of independently edited wiki projects such as *Wikipedia* was so surprising that its editors have sometimes referred to it as the "zeroeth law of *Wikipedia,*" stated as, "The problem with *Wikipedia* is that it only works in practice. In theory, it can never work."

Actually, it also works in theory; it's just that an emergent theory to incorporate its conduct has needed to emerge too.

CONFLICTS

Another scale-invariant power law exhibited by human activity is the incidences of conflicts. After seven years of research, in 1948 British mathematical physicist Lewis Richardson published an analysis of some three hundred violent encounters that had occurred during the previous century and that ranged from small-scale skirmishes to the two world wars.[4]

With this and further data on even greater numbers of conflicts,[5] what he revealed may seem extraordinary. He demonstrated that such displays of aggression, with their myriad possible causes and plethora of choices, obey exactly the same type of power law, in terms of the rigorously close and logarithmic relationship between their frequency and number of fatalities, as does the regularity and scale of earthquakes. This also shows that major wars aren't anomalous; tragically, they instead are extreme events on a continuous spectrum of human-caused disasters.

In recent years, researchers including Neil Johnson at the University of Miami have discovered that the same power law also applied to the occurrences of insurgent attacks against US forces in Iraq and Afghanistan. In widening their analysis to include the timing as well as the ferocity of attacks, in 2011 Johnson and his team presented a further paper outlining a method for predicting the evolutionary progress of conflicts. It shows a familiar relationship known as a progress curve, which tracks how productivity increases across a wide range of human activities through an iterative process of experience and adaptation as the continuing accumulation of information enables practice to make perfect.[6] Sadly, such "productivity" can occur in wars too as the two sides learn about their enemy and adapt accordingly.

The team's study found a progress curve that relates the logarithm of the attack number (first, second, third, and so forth) and the attack interval. By knowing the initial interval between the first and second attacks, their premise is that future assaults may be predictable.

What's perhaps even more revealing is that the adaptability of both sides reflects what's called a Red Queen process. In Lewis Carroll's *Through the Looking Glass,* the futile chase between Alice and the Red Queen eventually takes them back to where they started. In evolutionary biology the term is used to describe the competition between hosts and parasites or predators and prey, where the adaptation by one results in a corresponding response by the other that in due course leads to equilibrium.

What Johnson's team has effectively uncovered is the realization that the characteristics of recent, predominantly insurgent-type conflicts have and will likely continue to mire combatants in unwinnable and interminable Red Queen standoffs unless some form of sustainable resolution is found and undertaken to break the eventual impasse.

IT'S A SMALL WORLD

A further phenomenon that is innately holographic and demonstrated throughout human society is that of small-world networks. For three centuries, scientific research focused on taking stuff apart to identify its smallest and most basic ingredients. While such a reductionist approach has been hugely successful, it's inherently limited as it neglects the underlying connectivity that the more recent and progressively holistic approach of the past few decades has increasingly revealed.

An example is what's called small-world networks, which describe numerous situations where only a few links connect node points with each other, thus allowing influences to spread easily. The typical number of steps between nodes grows proportionally to the logarithm of the number of nodes in the network, making such a network innately scale-free and holographic.

Understanding the linkages of such small worlds took a notable step forward with a well-known and seminal experiment conducted in the late 1960s by American psychologist Stanley Milgram. In the days before the Internet (which itself has small-world attributes), he decided to test anecdotal evidence of small-world connectivity by writing a series of letters to arbitrarily chosen recipients in widely separated areas of the United States. In these letters Milgram gave the name and profession of someone in Massachusetts. He asked the recipients to pass along the letters to anyone in their own social networks who might be able to help the letters eventually reach their correct Massachusetts destination. Those letters that did manage to reach the target showed that, on average, the chain from himself to the final recipient comprised six steps, leading to the widespread concept of six degrees of separation. More recently, the same experiment was conducted on the Internet, with e-mails substituting for Milgram's snail mail, but nonetheless the same level of connectivity was still found.

In 1998, mathematicians Duncan Watts and Steve Strogatz, both then of Cornell University, were the first to show that numerous networks exhibit small-world characteristics. By modeling lattices they discovered that taking a regular grid and simply replacing a few of its ordered connections with ones of differing lengths they created a hybrid that combined clustering with freedom of movement and that optimized the connectivity of the entire grid—in other words, they created a small world.[7]

As Watts wryly noted, acknowledging the enormously broad spread of small-world applications: "I think I've been contacted by someone from just about every field outside of English literature. I've had letters from mathematicians, physicists, biochemists, neurophysiologists, epidemiologists, economists, sociologists; from people in marketing, information systems, civil engineering, and from a business enterprise that uses the concept of the small world for networking purposes on the Internet."

Small worlds enable and optimize information to flow highly efficiently throughout the network, even when it's limited to merely local knowledge. Far earlier than the small worlds of the Internet and social networks such as Twitter and Facebook, the prevalence of such scale-free connectivity throughout human societies engendered the saying of news spreading like wildfire—appropriate given that both neurons in the brain transfer data and forest fires do indeed spread as small-world networks.

E-MAILS AND LIBRARY BOOKS

You might think that researchers have more important stuff to focus on, but studies of web browsing, e-mail activity, and the borrowing of library books are further examples of the inherently holographic nature of our collective human behavior.

In 2005, Albert-Lászlo Barabási and Joäo Oliveira broke new ground in the investigation of human dynamics by looking at the timing of a number of activities in everyday life and work, including e-mails and snail-mail communications. Once again, their findings defied expectations. Instead of indiscriminate conduct they revealed, just as for earthquakes and conflicts, power-law structures of intermittent high bursts of action interspersed with low-frequency behavior.[8]

The following year Barabási and other co-researchers spotted power-law relationships in the visitation patterns of web browsing by analyzing the time intervals between consecutive visits by the same users to the site of a major online news portal.[9]

Further exploration of human interactions was undertaken by a group of researchers and published in the *Proceedings of the National Academy of Sciences of the United States* in 2009. Reviewing the communications patterns in two social Internet communities, they found them to again follow scaling laws between the fluctuations in the number of messages sent by members and their level of activity and revealing long-term patterns that stretched from days to more than a year.[10] Once again the findings surprised the team.

In 2010, researchers Chao Fan, Jin-Li Guo, and Yi-long Zha decided to study another common activity—the amount of library book loans over time—and again found the pattern of the frequency of their usage to be fractal. Converting the analysis into complex networks of actions, they also discovered them to be scale free and to embody small-world features.[11]

PREDICTABILITY AND CONTROL?

To see what patterns might emerge in terms of the physical movements of people, in 2010 Chaoming Song, Zehua Qu, and Nicholas Blumm joined Barabási to track the mobility of fifty thousand anonymized mobile-phone users over a three-month period. Anticipating that they would see a low level of predictability of such movements given the wide variations in demographics, peoples' differing arrangements between being at home and being at work, and distances traveled, they too were shocked to find a very high, 93 percent, level of predictability. As they noted in their results, given the diversity of the people they tracked, where they lived, and the lifestyles they followed, "Despite the significant differences in the travel patterns, we find a remarkable lack of variability in predictability, which is largely independent of the distance users cover on a regular basis."[12]

In recent years the exponentially expanding ability to mine and process vast amounts of data is enabling scientists to ever more quantitatively analyze human dynamics. Progressively, systems theorists are both utilizing this ability and also reframing an increasingly broad range of social and economic systems as complex networks. Focusing on their informational interactions following holographic power-law and thus scale-free patterning, they're aiming to determine the parameters of such interactions and maximize their levels of predictability.

While such research is still in its early stages, one important instance for our intermeshed globe is to understand and minimize the spread of disease. In late spring 2009, an outbreak of the H1N1 influenza virus was generally predicted to peak the following January, and, accordingly, fast-track development of a vaccine was planned to be completed in November.

Alessandro Vespignani and his colleagues, then at the University of Indiana, thought differently. Using complex network theory, they instead forecast that the outbreak would peak in October, and the vaccine would therefore be too late. As it turned out, while the virus wasn't as virulent as expected, the Indiana team's predictions were correct.

While thankfully the epidemic wasn't as severe as threatened, its relative mildness actually held off a comprehensive adoption of the new method of prediction. It took a much greater threat—the 2014 outbreak of Ebola—to do so. In July of that year complex-systems scientists at the Santa Fe Institute were invited to join a multidisciplinary group to mathematically model the epidemic as it tore through West Africa. Collating the various factors that affect disease transmissions to understand how their interactions cause a major emergency and so help determine effective interventions, they produced new insights—significantly, the link between poverty and Ebola. The countries worst affected—Sierra Leone, Liberia, and Guinea—have very few doctors and minimal health-care capacity. While neighboring countries only have modestly more health infrastructure, they nonetheless passed a critical threshold of capability to withstand the outbreak by isolating cases and tracing contact to those infected to then quarantine exposed individuals. While the response of the international community in providing additional health-care support during the 2014 epidemic was initially too slow, it then significantly ramped up (notably as foreigners began to catch the disease) and was finally contained.

Reducing poverty and improving health provision in poorer countries is a clear and present strategic need in order to prevent and manage future outbreaks. Nonetheless, the acumen provided by understanding the complex networks of interactions of those affected and the critical tipping points for deploying effective resources and support also provides key input to decision-makers.

So much information is now known about personal preferences, choices, and movements that researchers are becoming able to track the informational entropies, the dynamic informational content of their choices and actions, of every individual within a complex network.

This tracking and analyses of information relating to individual and collective choices and actions is indeed improving predictability, whether from the relatively trivial (but often annoying, at least to me) pop-ups on my computer that suggest future purchases from past choices to the hoped-for and benign ability to head off health and financial disasters and the co-creation of increasingly integrated social support.

Yet, there is other and more malign potential. We're already having the difficult debate between the perceived need for high levels of governmental surveillance with the aim of preventing terrorist attacks and the maintenance of personal privacy. We're having, though, far less discussion about how corporations such as Google and Facebook, in their pursuit of profits, are transforming our notions of what's private. Indeed, the interception of unencrypted broadcast traffic from mobile phones, for example, is legal in many countries and can be easily used by companies to build up progressively detailed personal profiles for profit-motivated purposes.

Data mining, however, and tracking the informational entropies of people have wider potentially adverse outcomes than a loss of personal privacy, which surveys have shown to be more of a generational concern, with younger people being apparently less bothered by such considerations. The increasing lack of privacy inevitably incurs reduced personal security as others know ever more about our behaviors and movements. Additionally, personal, corporate, and governmental systemized securities are also increasingly vulnerable to cyber hacking; there are a plethora of motives for such intrusions, but they are rarely benevolent. Enormously and rapidly increasing levels of information about us are, though, not only leading to less privacy and greater insecurity but also unavoidably to greater influence over us, whether by governments or corporations, which in turn can be extended to control. While we may take different views of their motivations—be they governments, institutions, or corporations—we might want to individually and collectively consider where we may be heading or indeed subtly (or not so subtly) being herded, before we reach points of no return.

NUMERICAL HARMONICS

As we've seen, the in-formational patterns derived from complex numbers are universal in terms of describing the underpinnings of physical reality. So-called natural numbers (one, two, three, four, etc.) aren't complex, yet are equally universal: discovered rather than invented.

Physicist Frank Benford and linguist George Zipf have laws named after them that, in rather unexpected ways, predict and describe the underlying harmonics of creation.

Benford's law states that the *relative* frequencies of the numbers one through nine follow a simple harmonic rule regardless of what phenomenon, system, or process they describe, at whatever scale and regardless of the unit of measurement. Instead of each of the numbers being represented equally as leading digits in large data sets, the number one is six times more likely to be so than the number nine. From diverse examples such as street addresses, the distribution of twitter users by number of followers to the numbers expressed in mathematical and physical constants, only two basic requirements are needed for its harmonic relationships to emerge. The first is that the sample of numbers that quantify the phenomenon is large enough to enable proportions to be established, and the second is that within the phenomenon the range of numbers is unconstrained.

The law is most accurately reflected when the data are spread across multiple orders of magnitude and can be found in numerous phenomena that obey the logarithmic relationships of power laws.

In 2010, mathematical geophysicist Malcolm Sambridge and his colleagues at the Australian National University in Canberra reviewed the presence of the law in fifteen different data sets across physics, astronomy, geophysics, chemistry, engineering, and mathematics. From examples that included the depths of earthquakes, the brightness of cosmic gamma rays reaching Earth, greenhouse gas emissions by country, and cases of global infectious diseases, they attested to its widespread applicability.[13]

Again, though, it also pervades phenomena and data arising from our personal and collective choices and shows up in places as diverse as population figures, corporate sales and costs, electricity bills, stock prices, and even the range of numbers collated from the pages of newspapers. So omnipresent is Benford's law that Sambridge proposed it as a new way of detecting anomalous signals.

In human behavior and transactions, its absence had been used to detect irregularities including evidence of financial fraud, and in 2013, Thomas Hair at Florida Gulf Coast University decided to go further afield to use it to test for such anomalies in the search for exoplanets. The database of the number of those already confirmed and the list of possible candidates were compared. When measured in multiples of either the masses of Earth or Jupiter, Hair found that most of the candidates do indeed adhere to the law, indicating that some 90 percent will eventually be verified.[14]

In a final elegant revelation, we even find this harmonic law of numbers interwoven with phi, for the digits that make up the Fibonacci sequence progressively conform to Benford's law.

Zipf's law, which also pertains to both "natural" and "man-made" phenomena was initially found to denote how the occurrence of any word is inversely proportional to its rank in the frequency table. So, astonishingly, the most frequent word in *any* language will occur twice as often as the second most frequent, three times as often as the third most frequent, and so on.

However, the harmonic nature of its description of inverse proportionality has been discovered to extend to many other circumstances. So, for example, if the cities in a country are ranked according to size of population, they'll be in such twofold scaling. Thus, if the largest city in a country has a population of 1,000,000, the second city will have 500,000, and the third 250,000. In 1999 economist Xavier Gabaix analyzed the populations of cities in the United States, showing the adherence to Zipf's Law[15]

In January 2015 astrophysicists Henry Lin and Abraham Loeb at the Harvard-Smithsonian Center for Astrophysics in Cambridge, Massachusetts, using population density as the critical variable (people and stars, respectively), even found the same scaling model to be accurate for the growth of cities and the formation of galaxies.[16]

MUSIC

We've previously noted that we both see and hear the world logarithmically. So perhaps it's not too surprising that spectral analysis of a wide range of natural sounds, such as waterfalls and waves on a beach and melodic music, both human made and birdsong, also reflect the holographically fractal distribution of so-called pink noise. Its embodiment of fractal features where the intensity of the sound diminishes with frequency to yield approximately the same energy per octave enables pink noise to feel naturally harmonious and resonant with our innate sense of rhythm.

As science progressively reasserts the ancient understanding of the coherence of the harmonic and holographic nature of the manifest world, such fractal attributes of sound and music are not only providing the sound track to our experiences, but their importance is being increasingly recognized in nonintrusive healing modalities that utilize such universal harmonics and resonances as well.

In 2014, researchers Weyland Cheng and Peter Law at the University of Science and Technology in Wuhan, China, and Hon Kwan and Richard Cheng at the University of Toronto in Canada reported on the use of music and ultra-sound (along with other modalities including the application of electromagnetic fields) in stimulation therapies for the treatment of illnesses.[17] Recognizing that healthy human biology is inherently fractal and that disease disrupts such patterning, they showed how the difference in dynamics between healthy and dis-eased physiology can be observed in many biological signals such as neural activity and variations in heart rate and breathing patterns.

By showing how optimal stimulation techniques are able to guide dis-ease back to health, they emphasize the value of pink sound and music to literally and effectively retune the human body and mind. They report a trend of increasing evidence from numerous studies that show such auditory stimulation to have positive health benefits for a range of conditions including depression, autism, and dementia.

BLACK SWANS

While the harmonics of power laws and complex systems often apply, as we've seen, across many orders of magnitude, they also demonstrate what's been dubbed by Lebanese American scholar Nassim Taleb as black-swan behaviors, where an unexpected and rare event occurs that has a major effect.[18]

Such disproportionally important happenings, throughout history, science, and finance, in hindsight are usually acknowledged as having been possible. Being hard to predict, though, and often beyond previous experience, they're often ignored or peripheralized until they actually occur and are then retrospectively rationalized. Examples quoted by Taleb include major scientific discoveries and the attacks of September 11, 2001.

Taleb took the metaphor of black-swan events from the Roman poet Juvenal's depiction of something being "as rare as a black swan," and which was used much later as a common expression in sixteenth-century London when it was presumed that black swans didn't exist, when in fact they actually did, albeit on the other side of the world. He uses the metaphor to reveal the weakness of such limited thinking, as the sighting of even one black swan undoes the viewpoint that mistakes the absence of evidence for evidence of absence.

Also citing World War I and the rise of the Internet as examples, among many others, of black-swan events, he summarizes them as being rare, having extreme effects, and being retrospectively predictable. Recognizing the difficulty in forecasting them, however, he rather urges robustness, what he calls strategies of antifragility and attempts to mitigate exposure to those with potential adverse effects (such as financial system collapse) and optimization of organization and resources to gain the full benefit of those likely to be beneficial (such as the rise of the Internet).

The ongoing challenge though, arising from their inherently nonlinear nature, isn't only recognizing them before they occur but the playing out of their evolving and often equally nonlinear consequences in the days, months, and years that follow. Just as in the past, future black swans, both positive and negative, *will* occur. Indeed it is the ones that would be catastrophic and for which we should seek ways to ameliorate their potential destructiveness that we're least prepared for.

Taleb and others have recognized the psychological barriers that often hold us individually and collectively in sometimes willful ignorance and denial and prevent effective preparation for such threats. Such denial usually masks deep fears, and if we as a progressively interconnected global society are to come together to plan for and meet such threats and deal with their aftermaths, we also need psychologically to not only recognize but also overcome such anxieties and the fear-based behaviors they engender.

Only then can we attain the clear-sightedness, flexibility, robustness, and effectiveness required to confront and deal with their rare but potentially cataclysmic inevitability.

ALL PHYSICAL REALITY

The challenge of answering philosopher and cognitive scientist David Chalmers's famously "hard" question about the nature of consciousness—how something immaterial can arise from something material (the brain)—is that it comes from an incorrect premise. Its fallacy is the assumed duality between the apparent immateriality of mind and the seeming materialism of the physical world. As we've seen throughout this book, as leading-edge science is coming to realize, such apparent separation is illusory. Instead the entirety of the physical world is being discovered to be literally the all-pervasive expression of informational processes.

That includes us. Our personal thoughts, emotions, choices, actions, and behavior patterns are distinctive. Yet, as we've considered, as more and more analyses of human activities are investigated, it's becoming increasingly clear that while arising from myriad individual decisions, our group and collective conducts embody exactly the same holographic signatures as are exhibited throughout the so-called natural world.

So, this is the point in our journey, when, if not already, there's now the need to acknowledge, confront, deal with, or embrace (depending on how you're personally feeling) the realization that *all* we call physical reality is expressed as a cosmic hologram, that each of us is a holographic microcosm, and our collective human experiences are a holographic "meso-cosm" of the macrocosmic in-formational intelligence that articulates itself as our Universe.

This recognition, though, raises another question: Who makes our perfect Universe?

PART 3

Co-creating in the Cosmic Hologram

11
Who Makes Our Perfect Universe?

An in-formed universe requires an in-former . . .

Look up at the stars and not down at your feet. Try to make sense of what you see, and wonder what makes the universe exist. Be curious.

Stephen Hawking, physicist

Many pioneering scientists, including Planck, Heisenberg, Schrödinger, and Einstein, have come to stand alongside spiritual seekers from all eras to gaze into the as yet unknown and ask the question of who or what creates our perfect Universe.

Indeed, from a letter he wrote in 1936 to young student Phyllis Wright, Einstein's viewpoint holds that "everyone who is seriously involved in the pursuit of science becomes convinced that a spirit is manifest in the laws of the Universe—a spirit vastly superior to that of man, and one in the face of which we with our modest powers must feel humble." I would maintain that we are now at a threshold where the next steps in revealing the essence of this great mystery are at last capable of integrating evidence-based and faith-based perspectives.

The growing scientific perception of the informational nature of all that we call physical reality is, at the same time, showing it to be in-formational, literally in-forming in its entirety and encompassing all from its simplest to its most complex forms.

In other words, our Universe is composed not from the all-pervasive presence of merely arbitrarily accumulated data and accidental processes, but ordered, patterned, relational, meaningful, and intelligible in-formation, exquisitely balanced, incredibly co-creative, staggeringly powerful, and yet fundamentally simple.

The in-formation present from the very beginning of space and time, as simple as it could be but no simpler, provided the essential instructions from which our 13.8-billion-year-old Universe has facilitated the evolution of ever greater levels of complexity. The informationally entropic progressions of its nonlocally connected intelligence has and continues to co-creatively express, explore, and experience on all scales of physical existence, as it advances to the embodiment of progressive self-awareness.

As there cannot be an in-formed universe without the existence of an in-former, without anthropomorphizing such creative impetus, the inevitable question that these scientific revelations come to is: Who or what is the ultimate intelligence that makes our perfect Universe?

We'll now pose this essential query and see how the emerging understanding of the cosmic hologram offers a visionary perspective and new insights, both on this age-old enigma and in answer to it. With the increasing evidence that ours is a finite Universe within an ultimately infinite cosmic plenum, it's also now an appropriate time for us to address notions of a multiverse of other universes beyond our own.

OUR IN-FORMED UNIVERSE

The French philosopher and author Marcel Proust once stated: "The real voyage of discovery consists of not seeking new landscapes but having new eyes." So has it been throughout the history of humanity, as we've come repeatedly to expand our vision of the world and see it and ourselves anew. With new data, we can amass further information and perceive the meaningful in-formation it encodes. From its intelligence we can progress to knowledge and onward to incorporate its strands into an interweaving tapestry of unfolding wisdom.

We began our exploration of the cosmic hologram by referring to the poetic and profound vision of the ancient Vedic sages of Indra's net. Describing the holographic expression of the infinity of cosmic mind, we find ourselves, some three millennia and countless discoveries later, coming to essentially the same perspective, albeit recounting it in very different language. This developing convergence of views continues with an acknowledgment of the innate and all-pervasive intelligence that flows throughout and indeed makes up everything that we call the physical world.

Our "univer-soul" journey over the past 13.8 billion years has progressed from its radiant beginnings to the geospheres of planets, on to the emergent and complex biosphere we are members of on our own planetary home, to the globalized techno-sphere of the early twenty-first century.

To where now do we journey? What do we now need to see and with what new eyes?

The techno-sphere we currently inhabit and the Internet are enabling the enmeshing of our societies into a global whole; we have access to vast new levels of information that flow as never before. This techno-sphere connects us literally as a worldwide web, raising and reflecting both the commonality of our human experiences and their diverse expressions. At its perhaps most opportune, it offers us ways to appreciate, share, and enhance universal values while learning to celebrate and honor our differences. It also without precedent enables us to come together at moments of great inspiration and joy or of deep sorrow.

Revealing what has been, what is, and what may be, for good or ill and like no other resource before, it encourages and sometimes forces us to collectively view with new eyes. What we then can or are willing to comprehend, how we choose to respond, and to what degree we're prepared to participate in being the change, as Gandhi said, we wish to see in the world, is up to us.

In the years following the cataclysm of World War I, the concept of the noosphere, that derives from the Greek word *nous,* meaning "mind," was developed by Pierre Teilhard de Chardin, Ëdouard Le Roy, and Vladimir Vernadsky. In thinking about the future of humanity, these three visionaries saw ahead to a potential for the processes of increasing complexity to evolve from an environmental biosphere to such a collective and unifying human and essentially planetary consciousness.

Now, almost a century later, the emergence of a techno-sphere that they never envisaged may come to be appreciated as being a necessary transition in enabling the whole of humanity to see itself, ask the fundamental whys and why nots of our lives, and set them in a wider context of the nature of reality itself.

In the "never again" time after the global conflict of the First World War, Teilhard perceived such a noosphere as embodying the victory of love over the forces of fear. Indeed he considered that "love is the affinity which links and draws together the elements of the world. . . . Love, in fact, is the agent of universal synthesis."

New discoveries are increasingly providing support for Teilhard's universal synthesis in the innately whole world of the cosmic hologram. At the same time, the techno-sphere is revealing the peril of global breakdown unless we come together as a human family and act inclusively, justly, and compassionately toward each other and responsibly for our one and only planetary home.

It may and hopefully will be, but only if we so choose, that as we approach the centenary of the farsighted vision of the noosphere that we are able to overcome our fears and denial and finally cross our next evolutionary and critical threshold to embrace its awareness.

COSMIC MIND

Beyond the complex plane, in its investigation of nonphysical realms, science is only just beginning to catch up with the numinous experiencers of past millennia. Its discoveries in the years to come will almost certainly shake our perceptions, not just of the world "out there" but at far greater and more personal levels. For they're likely to directly encounter the presence of a grander and ultimately infinite and eternal intelligence of which we are microcosms—droplets of the great cosmic ocean, sparks of the great cosmic flame.

The ongoing revolution in our understanding of the true nature of what we call "physical" reality and its inherent insubstantiality was only the first step of our journey. The next challenge is in perceiving how cosmic mind constructs and real-izes the in-formational and holographically expressed nature of ours and the possibilities of other finite universes; this is the pioneering scientific work-in-progress of seeking to understand the cosmic hologram.

Along the way, the perceived separation between mind and matter falls away, and the illusory dualism that has been under increasing threat from scientific discoveries over the past century can finally be resolved into the emergent understanding of the all-pervasive unity and wholeness of all-encompassing mind.

REAL-IZING

Neuroscientists, psychologists, and psychiatrists are coming to recognize that we don't perceive a direct representation of "external" reality, but instead our senses and brains operate as a translation and integration service to our innate consciousness. What we think, feel, and believe, whether or not it's "true," significantly affects our notions of what's real.

The old adage "seeing is believing" is being turned around to become "believing is seeing" as numerous studies and experiments have shown that we literally "see" what we believe, that we see what we expect to see. Psychologists have demonstrated that when our attention is distracted we miss otherwise obvious events and co-create the realities we perceive—proclivities often expertly manipulated by mentalists such as the UK's Derren Brown.

A well-known (and jaw-dropping) example is the phenomenon of "change blindness," demonstrated in an experiment carried out by psychologists Daniel Simons and Daniel Levin in 1998.[1] Such reality myopia occurs when our attention is diverted. Researchers have concluded that change blindness is due to a lack of informational attention before and after the distraction, so the brain fills in the gaps and concludes that no change has occurred, even when one actually has.

Simons and Levin's experiment was conducted at Cornell University. Experimenters held a campus map and asked passersby for directions. After about fifteen seconds of giving directions, two further experimenters, together carrying a door, walked between the conversers, with the initial experimenter who'd been asking the way switching places with the one carrying the back of the door, who then took the first experimenter's place in receiving help from the passerby.

When the passerby completed giving directions, the experimenter explained that he was conducting a psychological study of how people pay attention and asked whether the passerby had noticed anything unusual when the two people carrying the door went by. If the response was no, the experimenter then asked whether the passerby had realized that he or she wasn't speaking with the same person who'd initially approached to ask for directions. More than half the passersby failed to notice that the person whom they had been speaking with had changed to another stranger in the middle of their conversation!

Follow-up experiments revealed that unless there's some sort of vested interest in an encounter—such as when the appearance of the individual is relevant—even radical change, as, for instance, switching people in mid-event, goes unnoticed. While such an extreme example reveals change blindness in more than half of those tested, despite most people thinking that in less exceptional situations such levels of inattention wouldn't generally happen to them, it occurs for nearly eight out of ten individuals.

Scientists over the past couple of decades have also tested how what we think and believe actually alters our physiology. For Eastern practitioners of mind-based traditions such as meditation, such abilities have always been self-evident, but in recent years Western scientists have studied and increasingly corroborated such claims.

In 2002, the *Harvard Gazette* reported a series of experiments beginning in the 1980s and directed by Herbert Benson of the Harvard Medical School that investigated the Tibetan Buddhist meditative practice of *tum-mo.*[2] Not only were the monks involved in the experiments able to demonstrate a conscious ability to reduce their metabolism by up to 64 percent but also to significantly raise their body temperatures. Translated as "inner fire," tum-mo's visualization and focusing technique has been used for many centuries as a rite of passage to prove a monk's level of adepthood. The practice involves using concentration to produce sufficient inner heat to endure a nightlong meditation in the freezing cold, sitting naked on the snow.

With similar spiritual practices honed over millennia, Hindu yogis are also now being subjected to scientific investigation of their renowned powers to control bodily pain through the power of thought. In 2004 a team of researchers, with lead author Erik Peper of San Francisco State University, reported their study of a yoga master able to pierce his tongue and neck with skewers while suffering neither pain nor bleeding. Experimental measures showed that a coherently high level of alpha brain waves, associated with deep yet alert relaxation, revealed his ability to consciously reduce the electrical activity in his skin, thus reducing pain response and blood flow.[3]

The same lowering of electrodermal activity and consequential suppression of pain had also been demonstrated in an earlier 1999 experiment of hypnotism by Vilfredo De Pascalis and others, whose subjects displayed the same physiological response, but this time to hypnotic suggestion rather than their own conscious volition.[4]

Beliefs can consciously or subconsciously have the same effect. In the West, the well-known placebo effect, where someone's expectation of health improvement due to a particular treatment or medicine, despite the placebo being neither treatment nor medicine, nonetheless leads to a real alleviation of symptoms, including pain relief.

A pioneering experiment in 1996 by psychologists Guy Montgomery and Irvin Kirsch at the University of Connecticut treated students with an ostensibly new topical painkiller they called *trivaricaine.*[5] Only containing water, iodine, and thyme oil, it had no active ingredients. Yet, when painted on one index finger of each student, and then having the index fingers of both hands squeezed in a vice, the students consistently reported significantly less pain in their "treated" fingers.

As such an experiment and many others since have shown, the placebo effect is not the result of actively positive thinking, but a genuine belief, albeit erroneous in itself, in the power of the treatment to cure. So, once again the mind is shown to be able to control the physical response of the body.

In contrast to the placebo's beneficial effects, the associated and equally real so-called nocebo effect results in negative responses. In studying the response by subjects to nocebo suggestions, physiologist Fabrizio Benedetti at the University of Turin Medical School, has shown a hormonal response by the pituitary and adrenal glands, both of which respond to threats to our body with a fight-or-flight response. Literally if the belief and consequential fear were sufficiently powerful, such a response could be lethal.

In 2014, Benedetti proposed taking more than one hundred students on an Alpine trip, warning one of them beforehand that such high altitude could cause migraines. By the time of the journey, he found his comment had spread to more than a quarter of the participants. Those who'd heard the gossip not only had the worst headaches, but an analysis of their saliva also showed a heightened response to low oxygen levels and an increase in enzymes associated with altitude sickness, beyond that demonstrated by the other students. Social "infection" had become physicalized.[6]

Such mind-body effects seem to be increasing. In a 2009 article, *Wired* magazine's Steve Silberman reported that in the United States from 2001 to 2006, the percentage of new drugs cut from development after phase 2 clinical trials, when they are first tested against the placebo effect, rose by 20 percent. In more comprehensive phase 3 trials, the number of potential drugs failing tests increased by 11 percent, mainly as a result of poor comparison against placebo effects.[7]

Despite continuing high levels of research, progressively fewer new medicines are being launched owing to their inadequate performance against placebos. Silberman also reported that if follow-up tests on drugs that have been on the market for many years were now carried out, some would also fail for the same reason. Two comprehensive analyses of antidepressant trials also revealed a significant increase in placebo response since the 1980s, with one estimating a doubling of placebo-effect size during that period.

Significantly, it's not that the old drugs are getting weaker, but that the placebo effect appears instead to be strengthening. This has been put down to a number of factors. Increasing levels of advertising and drug tests in developing countries heighten the levels of expectation of a drug's efficacy. The increasing use of psychological drugs, such as antidepressants, also tends to affect the same parts of the brain associated with emotions and beliefs.

While all of these aspects no doubt contribute, may it be that a further influence is that both individually and collectively our underlying state of consciousness is increasingly and more powerfully co-creating our realities? In which case, the informational interplay between our thoughts, emotions, and physiology may be exhibited by such a correlation with both placebo and nocebo effects.

As we've seen throughout this book, psychological researchers, alongside those in the physical sciences, are also converging on the understanding that there's no "real" separation between a so-called observer and that which is observed.

Nonetheless, while there's an increasing appreciation that consciousness is primary to and co-creates our realities, as yet few leading-edge scientists have been willing to engage, other than philosophically, with the notion that *all* reality literally is consciousness at work and at play.

Neuroscientists and psychiatrists Giulio Tononi and Christof Koch are probably closest to a breakthrough in understanding the primacy of consciousness with Tononi's theory of integrated information, whose level of coherence is greatest during waking awareness, breaks down during deep sleep, and returns again to a high level during dream or REM sleep.

It currently, though, remains a big hurdle for many scholars to come to the recognition that our bodily senses and brains process and interpret the continual communication of the "I" (or indeed, "eye") of our individuated microcosmic consciousness with the "we" of our mesocosmic collective human and planetary awareness and the holographic entirety of our Universe.

BEYOND THE BRAIN

Realizing that everything we call physical reality *is* the expression of the in-formational intelligence of cosmic mind completely reframes the question of human and indeed all consciousness and cognizance. The view of most neuroscientists has been that our consciousness somehow arises from the brain as a localized phenomenon. Just as a turbine generates energy, so the view goes, our brain generates consciousness—somehow.

That final "somehow" is crucial. While neuroscience has managed to map neural networks in the brain, identifying which areas light up during certain mental processes and progressively noting their holographic nature, there remains no mechanism for the integration of neuronal activities with the immaterial perception of self-awareness.

Based on numerous experimental data on the nonlocality of our consciousness, an alternative has been to view the brain instead as a computer: a receiver and transmitter of nonlocal information. However, with increasing discoveries and the emerging understanding of the cosmic hologram, the computer-brain-mind metaphor is also coming to be seen as too limited in its perspective. Not only does it fail to account for the inter-dimensional communications reported extensively by transpersonal experiencers, but—and even more fundamentally—it still implicitly suggests a duality between the physical world and consciousness.

Instead, the paradigm-shifting view of the cosmic hologram, which recognizes the actual immateriality of the physical realm and the ultimate unity of consciousness, is offering a new view of the brain and its purpose. By identifying the brain as playing an important role in the in-formational *organization* of the embodied awareness of human beings, it redefines each of us as a unique and microcosmic individuation of the intelligence of the cosmic hologram of our Universe, literally making us co-creators of reality.

EXISTENCE, EXPERIENCE, AND EVOLUTION

Following Russian theater director Constantin Stanislavski's lead, acting coach Lee Strasberg introduced the idea of method acting to the United States in the 1920s. Such "method" immerses an actor in the personal identification of the role he or she is playing, often remaining "in character" not only when onstage or on a film set, but also beyond.

As individuated consciousness, when we incarnate on Earth, most human beings also undertake a process of such acting, where, as spiritual method actors, we immerse ourselves in the drama of our lives. By identifying with the roles we play, and, as we've seen, the erroneous perception of materiality, we may marginalize or even forget our innate spiritual consciousness. The egos of our self-awareness instead immerse us in the perception that we're separate, both from each other and the world around us.

The personas we inhabit during our lives are thus the lenses through which we perceive the Universe—how we relate to it and how it relates to us. So such method acting offers an intensity of presence, without which our associated experiences could otherwise appear less "genuine" to us, and in the sense of the cosmic hologram, less creative in its universal manifestation of the nature of physicalized reality.

By being embodied within the holographic harmony of our Universe we are literally expressions of its creative learning. Through its inherent relativity, reflections, resonances, and ultimate resolutions, we microcosmically express its existence, experience, evolution, and emergence. Such a recognition not only reconciles but also integrates the space-time experience of cause and effect with the spontaneous awareness of nonlocal consciousness that transcends physical reality.

Within space-time, the informationally entropic flow of time enables the choices and consequences of the life-learning expression of universal consciousness to progressively ensue. While, as an innately integral and nonlocally connected entity whose consciousness ultimately transcends space-time, the experiential co-creativity of our finite Universe in-forms itself and the infinite plenum of cosmic mind. This majestic view of ourselves and our Universe reflecting that of ancient spiritual wisdom, offers us an expanded and newly articulated vison of reality, and our own roles in its unfolding, as being fundamentally purposeful and meaningful.

THE WRONG QUESTION

We've arrived at a point where, given the ubiquity of consciousness, it seems clear that asking who or what makes our Universe is the wrong question. For progressively, our scientifically based understanding is showing that there's no "real" separation between maker and what's made; the appearance of such division is solely the perspective from which individuated aspects of consciousness may view their own holographic and holarchic imagery.

The eternal intelligence of cosmic mind finds a finite expression through the dynamic co-creativity of our Universe. Its existence, experiences, evolution, and eventual demise is as a unified and nonlocally cognizant entity, one of the infinite such expressions of the Oneness of all that is, has ever been, and ever will be.

Thus, each of us is a microcosmic and unique expression playing our own co-creative role in the unfolding self-awareness of the consciousness of our finite Universe and ultimately of the infinite Cosmos. While this has historically been seen solely as a religious, or more correctly, a spiritual perception, this limitation is no longer the case. Instead, the emergent viewpoint of the cosmic hologram that scientific discoveries are increasingly revealing arrives at this same conclusion.

God isn't "out there," a creator of the Universe and its creations. Instead, the greatest breakthrough we may make as human beings in the twenty-first century is to recognize that we and everything that we call reality in all dimensions and realms of existence *are* God, or whatever term we choose for the infinity of cosmic mind, and that we are microcosmic co-creators of its ineffable reality.

BIRTHS AND DEATHS OF FINITE UNIVERSES AND THE INFINITE AND ETERNAL COSMOS

The increasing acknowledgment of our Universe being finite yet the cosmic mind of the entire Cosmos being infinite and eternal, inevitably also brings us to the consideration of some form of a so-called multiverse within it, made up of other finite Universes that are also birthed, live, and eventually die: bubbles of thought in the infinite and eternal cosmic plenum. While it could be argued that all concepts of possible multiverse scenarios are still in the philosophic and prescientific stages of development, a few clues and pointers are helping to focus our attention.

To begin with, let's briefly review once again the increasing evidence for the finite nature of our own and thus other universes within an infinite plenum. First, on a fundamental level, the laws of physics appear to only operate with finite measures, which is incompatible with a premise of infinite space-time. The finite beginning of space-time also logically implies a finite ending. So does the informationally entropic process that not only gives time its start point and arrow but literally constitutes our experience of time itself. Such entropy also correlates with spatial expansion and consequential temperature reduction, ultimately falling to absolute zero or thereabouts in a finite timescale and when the process of increasing informational entropy reaches a universal maximum and so ceases.

Observationally, as we've seen, a 2003 examination of the energy patterns in the CMB showed that longer wavelengths are missing—a cutoff that also denotes a finite universe. More indications of ours being a finite Universe continue to be discovered. In 2012 an international team of astronomers led by David Sobral at the University of Leiden in the Netherlands analyzed the presence of H-alpha photons emitted by hydrogen atoms in association with star formation. They showed that half the stars that have ever existed were formed by around nine billion years ago, that the rate of formation has been reducing ever since, and that some 95 percent of our Universe's stars have already been formed.[8]

In other words, there remains only enough hydrogen in galaxies for a further 5 percent of stars to still be born, suggesting a relatively short future timeline, of the order of perhaps only tens of billions of years, until all stars are burned out. When our finite Universe does come to an end, the circumstances of how it will do so remain as yet unknown.

Three main scenarios have been postulated: the so-called big crunch, the big rip, and the big freeze. As yet, there's no viable theoretical mechanism of any of the trio and insufficient evidence for which, if any, might be realistic; though the big freeze with its maximal thermodynamic entropy is deemed the most likely. None of the three really deal with the notion of time or informational entropy, and to date very few cosmologists are testing any of them, or indeed any other option, in these terms.

One cosmologist who is venturing into this territory, however, is one of the greatest thinkers of his generation. In 2010, Roger Penrose, together with Vahe Gurzadyan, put forward the idea of conformal cyclic cosmology (CCC) as a model.

Set within the framework of general relativity, CCC envisages a continuing and holomorphic series of universes, each expanding from finite and tiny beginnings to an end point that can be reinterpreted as the start of the next cycle. For each spatially rescaled (or conformal) cycle, called an *aeon,* photons of light behave in the same way, so for light there is no boundary between one iteration and the next. The matter of fermionic particles, however, must convert to such electromagnetic radiations as a prerequisite for the conformal cycles to continue.

By recognizing that "space" is essentially a geometrical construct, CCC essentially relegates it to an emergent phenomenon. Instead, it perceives the notion of time, and thus the flow of informational entropy, as more fundamental.

Currently, the model doesn't consider the entropic maximization of information (equating to an end point at or near absolute zero) and holographically tie that to its perception of universal cycles. If it were to be modified to do so, Penrose and Gurzadyan's aeons, instead of being merely cyclic, would more explicitly equate to universes evolving, with the experiences of each lifetime able to in-form the next and so perpetuate emergence on a grand scale.

By emphasizing the significance of time and informational entropy, while the CCC model views the Cosmos as eternal, it implicitly refers to its only representing itself as a single uni-verse undertaking such infinite iterations. An alternative perspective places greater emphasis on the spatial aspect of reality, envisaging multiple universes although without specifically addressing whether they are finite, and if so, how they meet their end.

Instead, it considers how they might be born: not from infinitely hot and infinitely dense singularities but from finite seed-points, and to follow this line of thought is revisiting an extension to general relativity known by the initialism ECKS after its proposers—Einstein, Élie Cartan, Tom Kibble, and Dennis Sciama.

ECKS theory expands the conservation of angular momentum in the presence of a gravitational field to the intrinsic spins of fundamental particles and couples this with a nonzero torsional effect. When first proposed by Cartan in the 1920s, it gained some traction. However, as its effects seemed at the time to be of no consequence, and its added complication made the equations of general relativity even more difficult to solve, it was marginalized. That is, until Kibble and Sciama independently took further looks at its premise in the 1960s.

The current interest in ECKS, or some modification of it to incorporate the Planck scale and informational attributes of the cosmic hologram, focuses on its insights in possibly determining what happens under extreme conditions of space-time, such as at the centers of black holes or at the beginning of the big breath.

The tiny linkage it posits between the torsion and intrinsic spin of particles at extremely high densities causes a spin-spin interaction that replaces the infinity of a gravitational singularity with a finite but minimum scale. In addition, at this miniscule level, it forms an Einstein-Rosen bridge, a so-called wormhole, budding from and transcending the existing space-time of a collapsing black hole to create and in-formationally seed the expansionary white hole birth—the minute but non-singularity start of the big breath of a new universe. According to cosmologist Nikodem Poplawski,[9] not only does the ECKS model provide a finite birth process for universes but also naturally explains the flatness and homogeneity of space without the need for any inflationary mechanism.

The intrinsic torsion involved in such a birth might imply that our Universe too is spinning. This may also provide a correlation to an apparently preferred axis to space, the so-called axis of evil, as so termed by its discoverers, Max Tegmark and his colleagues, and could tie in too with the hypothesized perspective of the holographic bound of our Universe being toroidal in shape.

Indeed, recent observations may begin to corroborate such an axis and universal shape. A 2011 study of the rotations of more than fifteen thousand galaxies, led by Michael Longo of the University of Michigan, discovered a counterclockwise or left-handed preference in the direction of their spins. While the excess is small, it's nonetheless significant.[10]

So while indeed there's still a lot of work to do, the clues are beginning to add up.

It may seem at first that the CCC model and the ECKS mechanism may offer two incompatible versions of a multiverse. While one focuses on time and the other on space, they both consider reality as being fundamentally informational in nature.

By combining their approaches, I would suggest that a multiverse scenario may emerge that considers *both* the experience of space and time and indeed their innate relationship to be recognized, while also perceiving the infinity and eternity of cosmic mind to real-ize itself.

Bringing together some version of the ECKS mechanism to bud new universes during the lifetimes of existing ones while the CCC notion that their cyclic end-of-lifetimes gives further existence in a new aeon, also reflects major Indian traditions, including Buddhism and Hinduism, of spiritual reincarnation. In doing so, it extends the concept of evolutionary processes way beyond biology and even beyond a single universe to the entire Cosmos.

Such a scenario essentially not only embodies the grandest perspective of the cosmic hologram but also echoes the ancient Vedic notion of creation, as described in the Rig Veda (10:129): innumerable individuated manifestations of consciousness coming into form and then continuing to evolve through countless iterations as the cosmic mind of the Cosmos continues to eternally explore and co-create with all aspects of its ineffable self.

12
Supernormal

Not supernatural or paranormal . . . term coined by Frederic W. H. Myers in 1882

All the powers in the universe are already ours. It is we who have put our hands over our eyes.

Swami Vivekananda, Hindu monk who was key in introducing yoga to the Western world

By understanding that *all* that we call reality, not only on the physical plane but also beyond, *is* consciousness exploring and experiencing itself on myriad levels, the cosmic hologram offers an all-encompassing model of the Cosmos. As the emergence of the cosmic hologram's wholeworld-view expands our perception, it reveals there's essentially nothing that's supernatural or paranormal. Instead, experiences of nonlocal awareness that are capable of transcending space-time, while nonetheless extraordinary, should come to be seen as innate abilities.

By liberating ourselves from a limiting perspective, we can begin to not only understand but also consciously utilize the expanded insights offered by our inherent supernormal capacities, not merely to empower and benefit ourselves and enhance our own well-being and wholeness, but to appreciate and serve the ultimate oneness of all life.

REALLY?

In 1995 the American Institutes for Research (AIR) reported back to the US Congress on a study they'd been asked to undertake to review government-sponsored investigations by the CIA into remote viewing. Often referred to as a parapsychological or psi aptitude, remote viewing is when someone is able to gain a mental impression of a distant object or scene.

The conclusions by the two primary reviewers differed. One, who'd previously been open to the reality of remote viewing, was convinced by the evidence and in favor of then focusing on how such psi phenomena works. The other, previously skeptical, remained unpersuaded. The oversight panel then elaborated a consensus that, while agreeing there was a statistically significant demonstration of the ability to perceive on such a nonlocal basis, there was disagreement as to whether it could be unambiguously attributed to psi or some unconfirmed experimental bias. Without clearly establishing the cause of the proven evidence, the panel also considered that, even if it did exist, the experiments did not identify the origins or nature of the phenomenon.

In other words, while the evidence was there, the members still disputed its validity, regardless of the lack of any proof of error or bias; they couldn't, in any case, understand how it *could* work.

Twenty years have gone by, and many more psi-related experiments undertaken. The debate, however, between proponents, skeptics, and opponents of such abilities—which in addition to remote viewing include phenomena such as remote intention, telepathy, and precognition—continues to rage.

Indeed despite much contrary evidence that such phenomena are real, skeptics of parapsychology often dismissively refer to it as a pseudoscience.

Throughout the history of scientific discovery, theory and experiment have progressed rather like my husband's and my dancing: sometimes one leads, sometimes the other, and occasionally we manage to be in step. Where one or the other lags behind or fails to keep up, the dance can slow down or head off in an unsatisfactory direction for a while.

Physics has been called the parent of the other sciences, as the dance of its discoveries is fundamental to our increasing understanding of the nature of reality itself. Its scientific children then tend to follow its lead—sooner or later. Its ongoing development means that physics, as we've seen, eventually encompasses new findings and apparent anomalies in an ever more expansive theoretical framework. While some may be harder won than others, such advances are intrinsic features of the scientific method and its primary aim: to comprehend the physical world and how it comes about.

In the early twentieth century physics underwent a paradigm shift from perceiving things in a staunchly materialistic and dualistic manner to models based on relational fields of influence and energetic interactions. In the twenty-first century, information theory is coming to stand alongside physics to expand and deepen our perception and reveal the even greater ephemerality of physical reality, the primacy, of in-formation and consciousness and indeed the ultimate unity, of cosmic mind.

Some psychological and social sciences, especially those pertaining to human awareness, have barely caught up with the quantum and relativistic worldview, already a century old. While popularizing pioneers such as Fritjof Capra, David Bohm, and Rupert Sheldrake, among others, have sought to expand scientific and public awareness, most scientists are hardly yet cognizant that an even more transformational scientific revolution is under way, one that will radically affect not only their investigations but their own personal awareness too.

Lack of adequate theoretical frameworks hasn't prevented, let alone forced physicists to stop experimenting or barred them from attempting to observe new or anomalous phenomena. For parapsychology though, the lack of an "adequate" framework has often been used as an unscalable barrier to research. A commonly stated objection to such studies, that while they may work in practice they do not work in theory, has never been applied to any other scientific endeavor. There are many examples of where this has been the case, including where the actual physical phenomena arising from the quantized nature of energy and matter and the relativistic nature of space and time has been utilized in numerous technologies, without hitherto the reconciliation of quantum and relativity theories—the two theoretical frameworks on which they are based. Although open-minded skepticism is not only healthy but indeed crucial to scientific progress, in the case of psi phenomena, the skepticism has too rarely been open-minded and is often vehemently oppositional, even hostile, frequently from a priori prejudice.

The materialistic dualism embraced by many such opponents is, however, being progressively shown not to be a viable worldview. Instead, the developing perspective about the cosmic hologram and the fundamental nature of consciousness is finally offering a theoretical context, a "plausible mechanism" within which to place and perhaps explain supernormal capabilities and occurrences.

So let's now see how the idea of the cosmic hologram offers such a basis and so begin to finally answer the implicit question posed by the AIR reviewers two decades ago: How can psi phenomena work?

There are a number of key points, all of which we've already explored but that are necessary for us to again bear in mind.

The first is the fundamentally in-formational nature of reality and the ultimate unity of consciousness.

The second is that human consciousness is not restricted to our brains or bodies.

Third, nonlocal connectivity, which transcends space-time, is innate to our entire Universe, enabling it to evolve as a single coherent entity.

Fourth is that *within* space-time, the entropic flow of information engenders both time's arrow, the flow of time itself, and the experiences of causes and consequential effects.

Fifth, between all energy-matter manifestations of a system exhibiting nonlocal behavior, there's no entropic transfer of information *within* space-time. Literally, the whole system behaves, and indeed, for the manifestation of its nonlocality, *is* integral—regardless of the apparent spatial or temporal separation of its manifested attributes. What one feature "knows," the entirety of the phenomenon spontaneously "knows."

Sixth, *before* observation/measurement, *all* possible states of a system are not only nonlocally connected but in physics-speak are also described as being in superposition.

Finally, it's *only* when such nonlocally entangled and superpositioned states are finally measured that they, to use physics jargon, "collapse" to a specific real-ization.

Importantly, on measurement, *unless such measurement is precisely set up to reveal their inherent nonlocal entanglement,* the superpositioned states indeed "collapse" to a specific (and so entropic and localized) state within space-time.

Taking all these points into account, we can see that supernormal capabilities are inherently nonlocal, involving informationally entangled perspectives of consciousness that transcend space-time. So, while our "everyday" experiences reflect a localized real-ization, the coherence of supernormality allows us to access our innate and indeed universal nonlocal awareness.

In terms of spatially related remote viewing, intention, or telepathy, a resonance occurs that creates a level of spontaneous correlation and conscious entanglement. As we'll later see, such supernormal phenomena are especially exhibited where there's a heightened intensity, whether of focused attention, emotional bonding, or some form of powerful image or occurrence.

The supernormal attributes, though, that raise the greatest hackles in opponents tend to be those that seem to involve the transcendence of time, such as future presentiment, precognition, and past retro-causation, with their *apparent* violations of the principle of cause and effect.

With regard to often cited claims for retro-causation, I feel a widespread misunderstanding exists. Initially arising from an incorrect interpretation of a version of the famous double-slit experiment in quantum physics, the error then became compounded by the wrong conclusions regarding what's known as delayed choice experiments.

In 2012, philosopher David Ellerman of the University of California explained how the mistake occurs through what he refers to as the separation fallacy, and he described why such experiments don't actually imply any causal violation.[1]

To explain, let's begin with the basic version of the double-slit experiment, which was originally used to show the superposed wave-particle nature of energy-matter. Most simply, this setup involves a beam of light being shone onto a flat plate in which are cut two parallel slits, and the light that passes through the slits is then viewed on a screen behind the plate. The wave nature of light causes it to go through both slits and then interfere to form a characteristic pattern of bright and dark bands on the screen. However, the light always appears to hit the screen at distinct points, as discrete particles, the varying bands reflecting the differing levels of particle impacts.

If two detectors are each placed a tiny distance beyond the two slits so that a particle ostensibly going through the second slit isn't able to reach the detector placed behind the first slit, the common interpretation is that a hit at the first detector shows the particle has to have gone through the first slit.

Ellerman maintains that this is where the fallacy of apparent separation enters the scene. Its error is the assumption that the quantum superposition is already in a specific state at the ostensible "separation" apparatus (the slit) rather than at the detector, where information is accessed by its measurement. Ellerman points out that, instead, for an incoming quantum entity the separation apparatus actually creates a nonlocally entangled superposition that continues beyond the slits and on until the measurement, when it "collapses" to a particular state. The separation fallacy mistakes the creation of the entanglement for a measurement.

Now it gets really interesting. A delayed-choice experiment envisages the sudden insertion or removal of the detectors, after the particle has entered the apparatus but before it would have time to reach them. The separation fallacy makes it appear as though by doing so, one can retro-cause a collapse or not at the point of the particle's entrance to the separation device (the slits).

Again, the fallacy is rectified by understanding that it's the separation apparatus (the slits) that actually creates an entangled superposition state of the alternatives that continue to evolve *until* a measurement is taken.

Given too that a detector can measure just one "collapsed" state, that specific state is determined by, and only at, the detector and crucially depends on how and what information is accessed. A further critical point is that the entangled superposition is maintained until the final measurement, which always occurs in the *present.* The future is never measured.

Nonlocality is informationally non-entropic. Just as there's no information transfer through space, there's no separation or information entropically passing through and indeed *being* the experience of time. Expanding Ellerman's insight, the superposition of the entangled consciousness involved in supernormal experiences is transcendental both spatially and temporally until the final measurement.

In the case of retro-*causality* there's no difference in the nonlocal perception of the "present" and the "past" *within* space-time. Information can only be accessed in the present when the superposition of entangled alternatives is measured and so "collapses." So there's no violation of cause and effect, and a past that's already revealed isn't changed.

While we'll review some other experiments shortly, the supernormal phenomenon of presentiment or precognition seems to corroborate that the informationally entropic process that *is* the flow of time does, however, differentiate between a past that's already happened and a future not yet physically manifest.

In such precognitive cases, nonlocal awareness appears able to access a potentialized future whose superphysical in-formation is "crystalizing" but not yet fully real-ized to form the present. While as yet there's little evidence to show how far ahead such potentialities may begin to assemble as in-formational attractors in the complex plane, they seem to essentially form the superphysical bow wave of space-time. This makes eminent sense from the perspective of the cosmic hologram and the superphysical and dynamic nature of the underlying in-formational patterns of physical reality. Again, and significantly for such precognitive awareness too, *within* space-time there is no violation of causality.

So if we're now starting to approach a cosmological perspective that offers a framework for supernormal aptitudes of human awareness, we soon hit the next hurdle that has hampered investigation of these aptitudes.

All authentic scientific research involves rigorous experimental methodologies and protocols that are capable of enabling and ensuring that experimental results can be validated and replicated. Even in the so-called hard sciences of physics and chemistry there are, though, inevitable variabilities in results.

In his *An Introduction to Experimental Physics,* Colin Cooke wrote, "We repeat a measurement a number of times, make our best endeavors to get the same value each time—and *fail*" (emphasis in original).[2] In the studies of more complex phenomena, variations in environmental parameters and circumstances and the behaviors of the phenomena being studied inescapably increase the difficulties of obtaining accurate, verifiable, and replicable data.

Due to its inherent complexity and diversity, all experiments involving human consciousness are assiduously challenged to be demonstrable and able to be repeated. So the generic approach in medical, psychological, and social sciences relies heavily on individual case studies and collective meta-analyses of broad data sets to statistically prove a certain result or trait.

For parapsychology, this challenge is heightened still further given that supernormal capabilities for the great majority of people are nonordinary and fleeting; this is hardly surprising, if, as I've proposed, a fundamental purpose of our persona-based consciousness is to immerse ourselves in the experiential "reality" of the physical world. While supernormal phenomena and abilities puncture our preconceived notions of physical reality and our usual perceptions of being separate—which as we've seen are illusory—they categorically do not contravene the principles that play out within space-time, especially the flow of informationally entropic time and the associated processes of cause and effect. In this sense, they're a subtle, yet ever-present reminder, if we care to see or listen, of the wholeness of all that we call reality and the intrinsic integrity of our perfect Universe.

At the same time that the cosmological model of the cosmic hologram is gaining traction and the unity of consciousness is being remembered within our collective psyche, perhaps it can only be thus that our innate supernormal capabilities are emerging from the hinterland of superstition, pseudoscience, and denigration into the scientifically dawning and clear light of day.

In our 2008 book *CosMos,* Ervin Laszlo and I gave an overview of the experimental evidence for supernormal faculties up to that point.[3] We noted how researchers had recognized the challenge of biases, had progressively gone to greater lengths to ensure robust methodologies, and had continued to produce a plethora of positive experimental evidence for psi.

One quote sticks in my mind from engineering physicist Robert Jahn, the head of the Princeton Engineering Anomalies Research (PEAR) lab, who with his colleague Brenda Dunne undertook nearly three decades of research and amassed an enormous archive of evidence for such phenomena. On his retirement in 2007 and when the lab was then closed, he said, referring to those who adamantly opposed their findings: "If people don't believe us after all the results we've produced, then they never will."[4]

Since then, even more studies and meta-analyses have continued to show that increased experimental controls haven't reduced the substantial support for the existence of such phenomena. In addition to earlier investigations, meta-analyses of extrasensory perception by psychologists Lance Storm, Patrizio Tressoldi, and Lorenzo Di Risio in 2010[5] and 2012,[6] and by Storm, Tressoldi, and statistician Jessica Utts in 2013[7] and Tressoldi in 2011[8]; of precognition by psychologist Julia Mossbridge, Tressoldi, and Utts in 2012;[9] and of remote intention by psychologist Stefan Schmidt in 2012,[10] all provide further proof.

NONLOCAL AWARENESS: BEYOND SPACE AND TIME

Let's now briefly review some of the types of explorations into supernormal phenomena that are being undertaken, what clues from earlier studies are being reinforced, and what additional insights are emerging. We'll begin with those that exhibit nonlocal entanglement that transcends spatial distance: remote viewing, remote intention, and telepathy.

Over decades of studies, the accuracy of remote viewing has been found to be enhanced when the viewer is in a relaxed yet alert mental state of non-expectancy. Generally, this is facilitated by quiet and non-distracting surroundings and under laboratory conditions sometimes involving mild sensory deprivation, often called the *ganzfeld* approach, intended to minimize the level of environmental "noise."

While it's called "viewing," research has shown that the awareness reported isn't always experienced visually. It may also be "heard," be felt in the body, or be generally impressionistic—an inner-sight rather than a mental knowing.

Cognitive scientist Stephan A. Schwartz (www.stephanaschwartz.com), a pioneer of remote-viewing research, and others have also discovered that its precision also increases in relation to the scale of informationally entropic gradients (essentially the dynamic intensity) of the images, objects, or occurrences that are being nonlocally perceived. Significantly, those that are more powerful, emotionally engaging, or depict sudden or substantial change elicit the strongest and most specific responses.

It seems that the combination of the quieted attention of the reviewer and the emotionally compelling nature of the target maximize the nonlocal superposition of the scope of awareness and shared information. Indeed, the yin-yang complementarity of the receptivity (whether cognitively aware, subconscious, or autonomous) by a supernormal experiencer and the intensity and dynamism of the informational content of the event, target, or circumstances the reviewer's consciousness is in nonlocal integration with, seem to be universal aspects of all supernormal phenomena.

Until its closure in 2007, when its extensive body of research was placed under the auspices of the International Consciousness Research Laboratories (ICRL), Princeton's PEAR lab had primarily studied the nature of human consciousness through its interaction with sensitive physical devices called random number generators, or RNGs.

The experiments attempted to harness remote intention to alter the output of such RNGs, changing their randomly generated data to measures that include a higher informational content. While the effects were generally small, over millions of trials carried out by several hundred operators, the PEAR archive amassed highly significant deviations from random expectations.

The results also demonstrated disparities between the performances of individuals over a period of time and between female and male operators. The lab also found that when two people telepathically worked together in remote-intention experiments, the results were measurably greater than either was able to effect alone. In addition, when there was an emotional bond between the two, the results were even stronger.

When monitoring group activities through the outputs of RNGs, such emotional attunement and coherence was especially noted. Group rituals and events at sacred sites, music concerts, and theatrical performances all generated larger RNG deviations, whereas mundane conferences and business meetings appeared to hardly engender a response at all (now there's a surprise!).

Telepathy is perhaps the most anecdotally reported supernormal phenomenon; it is especially common between people who share some form of close emotional connection. Such rapport has been widely studied in the case of twins, with one in five identical and one in ten fraternal siblings claiming such nonlocal perception. Extensive examples of such telepathic connection between twins has been reported by researcher Guy Lyon Playfair in his book *Twin Telepathy: The Psychic Connection.*[11] Again, the telepathic awareness tends to be at its strongest when one of the twins is in danger or undergoing an especially potent experience.

In 2013 a purported and especially powerful telepathic bond was investigated between a nine-year-old autistic Indian girl, Nandana Unnikrishnan, and her mother, Sandhya.[12] Tested and documented by a team comprising a psychologist, a social worker, and an educator and witnessed by other nursing staff, Sandhya, while in another room, away from her daughter's presence, was given a six-digit number and a poem to memorize; Nandana was subsequently able to recite both without any prompting. Once more, it's the close emotional bond that's significant, as Nandana doesn't appear to have a conscious telepathic connection with anyone else. As is the case with other supernormal phenomena, some form of resonance is an important factor; that is, being on the "same wavelength" is significant.

From the mid-1990s researcher Dean Radin of the Institute of Noetic Sciences has been undertaking a series of experiments investigating presentiment, which is a physiological rather than cognitive response to yet-to-occur stimuli. In the experiments, volunteer participants are seated in front of a computer screen that shows randomly generated visual images from a large database containing varying levels of emotional content; their electrodermal activity (EDA) levels—that is, the electrical conductivity of their skin—is monitored before, during, and after they've viewed the images. While unsurprisingly, graphic images produced higher intensities of EDA than visually calming images, what Radin has unexpectedly found is that such differences in response begin on the order of two seconds *before* the arbitrary images are flashed onto the screen.[13]

Throughout the series of experiments, incidences of such presentiment have consistently been observed and using the same methodologies have been replicated by a number of other researchers including Dick Bierman of the University of Amsterdam.[14] In 2004, a team at the HeartMath Research Center in California also found the same presentiment showing up physiologically, this time in measuring heart rate.[15]

Rather than unconscious presentiment, in 2011, social psychologist Daryl Bem of Cornell University submitted a paper focusing on precognition, the mental perception of future events which was peer reviewed and published in the *Journal of Personal and Social Psychology.*[16]

In his research Bem utilized a number of common and widely accepted psychology-experiment methodologies, but instead of running them forward in time, he ran them backward. One such generic experiment, proved in numerous instances, looks at the effects of subliminal messages on the responses to subsequent images, which is an effect known as priming. For example, the word *ugly* subliminally flashed on a screen results in responders taking longer to decide that an independently judged beautiful image, later shown them, is indeed so. Bem, instead, ran the test backward and found the same effect time-reversed, albeit much less pronounced.

Another common psychological trial has participants shown a series of words on a computer screen, and then later asked to type randomly chosen words from the series; the typed words, as would be expected, are better remembered during a subsequent memory test of the initial list. Again, Bem turned this backward. His participants were instead first shown a full list of words, then asked to recall them from memory, after which, words from the initial list were then randomly chosen by computer. The participants were shown to have been better at recalling those words from the initial list that would later be randomly chosen.

In all, his studies continued for eight years and involved more than a thousand students and nine different types of experiments, with eight of the nine demonstrating statistically significant precognitive effects. The furor the paper unleashed was virulent, despite Bem's undoubted authority in the field, his reputation as a meticulous researcher, the prestigious nature of the journal, and four peer reviewers passing the paper.

All relevant evidence is also showing that the acuity of nonlocal awareness, whether spatial or temporal, is substantially increased with the level of informational entropy, the dynamic informational content involved. As we've noted, researchers into remote viewing have found that the precision of responses increases in relation to the scale of informationally entropic gradients—that is, the intensity and dynamism within the images, objects, or occurrences that are being perceived. The more powerful, engaging, or transformational the target, the stronger and more specific is the reaction.

Radin's and Bierman's research into presentiment found the same correlation, with the response to powerful images being more prevalent. Premonitions, whether personal or collective, also tend to be far more associated with the precognition of dangers and disasters. On September 13, 2001, in the aftermath of the 9/11 tragedy, Sally Rhine Feather, the director of the Rhine Research Center in North Carolina founded by her father, the renowned parapsychologist J. B. Rhine, and a repository of precognitive accounts for decades, reported the largest number of premonitions ever received by the center in regard to a public calamity.[17]

Over hundreds of remote viewing sessions in the 1980s, Schwartz also discovered that where images involve some form of significant alteration, it's this element that is most accurately picked up.[18] This point has also been identified by physicist Edwin May, who, in a series of experiments building on earlier investigations, coauthored a 2000 paper with fellow researchers James Spottiswoode and Laura Faith in the *Journal of Scientific Exploration.* In it they too correlated the specificity of nonlocal viewing of images to the level of dynamism of their informational content.[19]

STILL ME

In 2014, an international research team sponsored by the University of Southampton in the UK reported on a study begun in 2008 of the near-death (NDEs) and out-of-body (OBEs) experiences of more than two thousand patients in fifteen hospitals in the UK, United States, and Austria.[20] This, so far the largest investigation of its type, aimed to determine whether claims of awareness during such experiences are real or hallucinations.

Their findings led lead author, medical physician Sam Parnia of the State University of New York, and his co-researchers to conclude that the experiences of death appear far broader than what is yet understood. They considered too that in some cases of cardiac arrest, which is synonymous with biological death, patients' memories of visual awareness are compatible with OBEs and may have corresponded to actual external events taking place during their cardiac arrests.

More than a third of those resuscitated following cardiac arrest and afterward able to be interviewed reported coherent awareness but without any specific recall of events, suggesting, according to Parnia, that there may have been mental activity subsequently lost either because of injury or sedative drugs. Two percent of those interviewed, though, exhibited full perception compatible with OBEs and with explicit "seeing" and "hearing" recollections.

One case was confirmed using auditory stimuli during cardiac arrest when consciousness appeared to be present during a three-minute period of no heartbeat, this despite brain function typically ceasing within thirty seconds of heartbeat cessation and not beginning again until the heart is resuscitated. In addition, the memories were detailed and consistent with verified happenings.

Their results, building on earlier smaller studies and many anecdotal incidents, have convinced Parnia and his co-researchers that "the recalled experience surrounding death merits a genuine investigation without prejudice."[21]

Irrespective of faith, beliefs, and even the insights of personal experiences as reported by numerous testimonies, the cosmological perception of the essential unity of consciousness as now exemplified by the cosmic hologram offers a scientific basis that frees our awareness from its apparently physical restrictions. The expansion of this developing perspective not only no longer forces our consciousness to be viewed as being restricted to a physical body but also liberates it from being tied to a physical lifetime.

A study in 2001 led by clinician Pim Van Lommel and colleagues at the Rijnstate Hospital in Arnhem, Holland, also investigated patients revived after the clinical death of cardiac arrest and their experiences of NDEs.[22] Their research of 344 cases concluded that these are authentic experiences, "which cannot be attributed to imagination, psychosis or oxygen deprivation."

One of the significant findings from their interviews with those who had consciously experienced an NDE was a commonly expressed profound sense of peace. Leading to a strong belief in an afterlife and newly felt freedom from the fear of death, such sentiments were not shared by those who had also undergone resuscitation but had no recollection of an NDE.

There are four other related areas of nonlocal consciousness research that have been plagued by even more controversy and that relate to the potential continuation of the psyche after the death of the physical body. These include investigation for evidence of such continuing awareness, the ability to communicate with this awareness, discovery of evidence for Earth-bound spirits or ghosts, and claims for reincarnation.

Despite an enormous amount of anecdotal evidence across cultures and over many centuries and more recent recording of numerous case studies, the materialistic paradigm of reductionist science has nothing but derision for such experiences. Regardless, however, of where future investigations may proceed—hopefully with greater open-mindedness allied to rigorous experimental procedures—the holographic worldview of the cosmic hologram and the essential monism of consciousness inherently encompass the possibilities for such phenomena.

EXTRAORDINARY ORDINARINESS

We shouldn't expect that supernormal attributes are other than extraordinary. Otherwise, our ordinary experiences of being human would be significantly compromised. The method acting that persuades us of our human persona, our experience of the richness and manifest complexity of the physical world, and the co-creativity of its evolution, would consequently be less convincing.

The issue, however, is that by treating such non-ordinary awareness as nonexistent, we've gone to the other extreme, persuading ourselves of the materiality and sole reality of the physical world. In doing so, the deeper meaning and purpose of our existence have been marginalized, and we have essentially separated ourselves from each other, from the wider Cosmos, and from our own expanded consciousness.

In 2014, psychologist Etzel Cardeña of Lund University in Sweden authored an open letter, cosigned by ninety-nine fellow academics[23] from Argentina, Australia, Brazil, Canada, mainland Europe, Iceland, Israel, Japan, Scandinavia, South Africa, the UK, and the United States, that called for an open, informed study of all aspects of consciousness, including supernormal traits such as telepathy, remote viewing, and precognition.

The letter lays out that contrary to the negative impression given by some critics, scientific research into parapsychology continues to be carried out in accredited academic institutions around the world despite its prejudicial treatment by some opponents, its taboo status, the subjection of its researchers to professional and personal attacks, and lack of funding.

It also points out that in the face of negative attitudes, scholarly and peer-reviewed articles supporting the validity of supernormal phenomena continue to be published in associated academic journals. Significant positive results, as we've seen, go on being reported, despite the antipathy by some mainstream scientists and reviewers that have meant that for many years some journals have ignored these while encouraging the publication of studies showing null and negative results.

Supernormal aspects of human consciousness are far too important to be marginalized, derided, deliberately misinterpreted, attacked unfairly, or dismissed out of hand. The results of ever-more strictly managed, comprehensively monitored, and well-analyzed experiments require acknowledgment.

As Cardeña and his colleagues state, "Dismissing empirical observations *a priori,* based solely on biases or theoretical assumptions, underlies a distrust of the ability of the scientific process to discuss and evaluate evidence on its own merits. The undersigned differ in the extent to which we are convinced that the case for psi phenomena has already been made, but not in our view of science as a non-dogmatic, open, critical but respectful process that requires thorough consideration of all evidence as well as skepticism toward both the assumptions we already hold and those that challenge them."

Given the emergent understanding set out in this book that unifies reality and naturally incorporates supernormal phenomena and experiences, it is surely time for scientists to answer the challenge of Cardeña and his colleagues and undertake a thorough and open-minded investigation of the various aspects of such psi manifestations, with their potential to expand and transform human awareness.

13
Co-creators

The action of and participation in the dynamic process of bringing all that we call reality into existence . . .

An individual can't create anything itself. All of our dreams come true with the cooperation and co-creation of other souls.

Hina Hashmi, author of *Your Life, A Practical Guide to Happiness, Peace and Fulfilment*

We've already arrived at the perspective that we're each individuated microcosms of the holographic intelligence of our Universe and ultimately of the infinite and eternal mind of the Cosmos. The Sanskrit word *maya* is often translated as meaning "illusory" and sometimes used in reference to the "illusion" of the physical world. While the apparent duality between mind and materiality is indeed illusory, an alternative interpretation of *maya* is "partial," which rather validates our human experiences and the reality of our Universe, yet sets them in a vastly greater context of multidimensional consciousness.

Having immersed ourselves in the partial worldview of maya materiality, the emerging perception of the cosmic hologram offers a holistic framework for us to expand our awareness and re-member the wholeness of who we *really* are. As we do so, we're empowered both personally and collectively to real-ize our meaningful and purposeful roles in a univer-soul perception where we are both creation and co-creators.

CO-CREATING OUR REALITIES

The most recent lab experiments shed light on a persistent enigma that's dogged quantum physicists and philosophers alike in the understanding of the nature of physical reality: the famous question of whether, if a tree falls in a distant forest with no one to hear it, does it make a sound.

As we've seen, the experimentally proven recognition that there's no reality until it's observed or measured would seem to infer that unless there's "someone" to hear it, the falling tree makes no sound. It can equally be argued that there's not a tree or even a forest without such observation. This conclusion, though, that there's not even a tree or a forest, takes an important one-step-forward insight but then a step sideways in that it effectively considers that the only conscious observer is a "someone" and presumably that someone is human.

However, the cosmic hologram reveals the unity of consciousness and its macrocosmic expression as our Universe, all-pervasive and essentially unified while being played out on all levels and scales of awareness. So, there's *always* "someone" interacting with the tree—microcosmically, meso-cosmically and macrocosmically. The falling tree is always witnessed.

Reality is thus co-created at all scales of existence and myriad levels of awareness. While we, individually as unique microcosms and collectively as meso-cosmic intelligence, contribute to the co-creative experience of our Universe, our human awareness is by no means the only kid on the block.

Each of us, though, is personally unique; there's never been, is, or will be another such persona. Even if humans were to be cloned, each individual would be far more than their identical DNA. Yet, we also share generalized personality and character traits and collectively interweave fundamental and universal human attributes with many that vary by culture, ethnicity, gender and sexual orientation, and environmental and societal circumstances.

We all, however, to some extent respond differently to ostensibly the same situations. The significant and growing scientific evidence that what we "believe" is what we "see" and that our beliefs, thoughts and emotions affect our physical states, is also revealing that to a great degree they affect how we behave and experience our lives. The beliefs that underpin the self-concepts of our personas don't even need to be authentic, nor do they need to be conscious, to substantially affect our view of ourselves and the world and so drive our behavior; we literally co-create our realities.

Developmental psychologists track how our personal sense of self is progressively developed from early in life and acts as the informational framework or skeleton for our embodiment of further beliefs and consequently our perception of experiences.

Important events seed specific beliefs in our minds about ourselves, other people, and the world around us. When especially traumatic these seed-points can also engender a more generic informational imprint relating to abandonment, abuse, or some other archetypal pattern, which may then be reinforced by further events. Unless resolved, such dysfunctional perceptions can continue to play out throughout our lifetime, influencing how we treat others, how we perceive others treat us, and how we treat ourselves.

Once seeds are sown, behavioral psychologists have shown how easily they're watered by further events and circumstances until they form aspects of a concretized belief system and self-concept that is often then difficult to change. Educational studies in the United States and elsewhere have revealed how primary-school level disaffection and lack of positive relationships between teachers and pupils and between schools and families lead to lower educational attainment and limiting perspectives very early, and which, if unresolved, then continue as an ongoing trend.

When such limiting beliefs are instilled, whether by educators or other authority figures, by family members or those of a peer group, or generally by society, they set a low bar of expectations. While some individuals rebel against such stereotyping, most acquiesce and so further self-limit their potential.

Conversely, a pioneering 1964 study by Harvard professor Robert Rosenthal showed the power of affirmative expectations.[1] The research involved teachers being told (wrongly) that a certain IQ test was able to predict the future intellectual prowess of their pupils. After the test, several children from each class were arbitrarily chosen and their teachers told (again erroneously) that the test had shown that these specific children were on the verge of great increases in aptitude. Over the following two years Rosenthal found that their teachers' expectations did indeed positively affect the subsequent IQs of the chosen children. More research showed that in numerous, often subconscious, ways the teachers expressed their positive expectations of these pupils, giving them more time and attention and offering them more encouraging verbal and nonverbal feedback.

Numerous experiments have also shown how powerfully our unconscious beliefs, presumptions, and stereotypical conclusions bias our decisions and actions, even when we're convinced of our fairness and lack of prejudice. Such evidence reveals that we're seldom truly open-minded but rather we progressively accumulate information in support of our subjective judgments. We rarely objectively collect information to form new beliefs, even when we consider that we're doing so. Instead, we generally gather and filter additional data to support and augment our already existing beliefs and biases.

Associated cognitive tendencies include so-called confirmation bias where we tend to agree with people who agree with us. As a result we're likely to surround ourselves, personally or via media, with those who have similar views, and we are inclined to dismiss those who disagree, regardless of any impartial evidence.

There are many such predispositions studied by psychologists that blind us to objective perspectives and that instead powerfully mold our worldview and experiences. What's especially problematic is that, as researcher Emily Pronin of Princeton University points out, "It's not that we're blind to the concept of bias, or to the fact that it exists. We're just blind to it in our own case."[2]

Since 2009, social psychologist Paul Piff at the University of California, in a series of laboratory experiments, has been researching the influence of beliefs, specifically in relation to the conduct of social hierarchies. One of his experiments, testing the behavioral effects of monetary inequality, involved more than a hundred pairs of volunteers playing a rigged version of monopoly.[3] The toss of a coin gave one player an arbitrary advantage over the other. Aiming to be reflective of real social consequences that commonly mean that having more money opens up more opportunities, increased access to greater resources, and disproportionate influence, the "rich" player was given a larger initial allocation of money, allowed to throw more dice, and to collect more money when passing Go than the "poor" player. Despite their awareness of being unfairly privileged, as the games progressed the "rich" players became ruder and less sensitive to their opponents and exhibited more domineering behavior. At the end, when asked what contributed to their successes, the "rich" players emphasized their own capabilities and how they'd "earned" their inevitable wins, rather than acknowledging the fixed setup.

In other experiments Piff and his team have shown how, as monetary wealth increases, while often unacknowledged or rationalized by their perpetrators, so does a sense of entitlement, selfish behavior, and incidences of lying and cheating in pursuit of goals while levels of compassion and empathy reduce. Yet, by using small psychological nudges to remind participants of the benefits of cooperation and consideration, they've shown mitigation of such tendencies, revealing that such beliefs and their consequential behaviors can be modified.

The more dogmatic our consciously and unconsciously held beliefs are, whether about ourselves, other people, the world at large, or the nature of reality itself, the more difficult it is to change our mind, regardless of the evidence provided. When we come across information that conflicts with our perception of reality, we may even suffer what's called cognitive dissonance, which is often psychologically painful. We tend to respond by denying, avoiding, or rationalizing away any evidence that offers contradictory views to the categorizations that make up our perceived values, truths, sense of identity, and notion of reality.

Both experiments and numerous case studies have shown that tension exacerbates both unconscious bias and cognitive dissonance. Such a stressful situation is encountered during what philosopher of science Thomas Kuhn termed paradigm shifts, where scientific progress undertakes a revolutionary leap in understanding that scientists would have previously considered invalid. Given that competing paradigms of the nature of reality often appear irreconcilable, such as reductionist materialism and the unity of consciousness, Kuhn also maintained that the comprehension of science can never be wholly "objective" but inevitably incorporates the subjectively perceived self-concepts and associated worldviews of the members of the scientific community.

It's no wonder then that many scientists have such personally painful cognitive dissonance with an emergent scientific revolution that not only transforms the understanding of the physical world but upends their personal views of themselves and the nature of reality itself.

EIGHT CO-CREATIVE PRINCIPLES

By recognizing that all we call reality is in-formational and essentially the wholeness of consciousness exploring and experiencing itself on many levels of existence, every principle of co-creativity in which we engage, whether consciously or subconsciously, must follow the same rules of informational physics through which the manifest reality of the cosmic hologram is expressed in our Universe.

I've previously written about an octave of eight such codes of co-creative perception, gleaned from the universally relevant insights of many spiritual traditions and my own researches and experiences over many years and that do indeed correlate with such informational precepts.

These eight signposts to wisdom are the principles of relativity, resolution, resonance, reflection, change, choice and consequence, conservation, and concession.

Here I'll just briefly summarize them.

The *principle of relativity* expresses the truism of John Donne's poem that "no man is an island, entire of itself." Everything in our Universe, from the fundamental relativity of space and time and the interweaving patterns of energy-matter, exists and is mediated through the polarities of real-ationships. All our experiences too are played out through encounters with different aspects of ourselves, our interfaces with others and with the wider world. Indeed, it's through the myriad interplays of such relativities that our entire perception is informed.

Donne's poem goes on to state that instead of apparent separation, "Every man is a piece of the continent, a part of the main." So, in transcending such relativities, the *principle of resolution* expresses their attained reconciliation, balance, and ultimate integration (or "intogreation"), where our illusory perception of duality is progressively re-soulved into an awareness of the unity of consciousness.

The harmonic and coherent relationships that pervade our Universe and that are the signature of the cosmic hologram manifest through the *principle of resonance,* which we individually and collectively embody on physical, emotional, and mental levels. We're "on the same wavelength" with people we like, or something "chimes" with us. Conversely when we're dissonant, we're literally out of tune.

Our well-being derives from our being in resonant and harmonious relationships of all types. Experiments have shown that when we live in dissonant states of chronic fear or anxiety, the continuing level of disharmony and stress negatively affects our mental, emotional, and physical health, weakening our immune system and causing such dis-ease as depression and circulatory and stomach and bowel problems.

When, though, we're positively in tune with our surroundings and circumstances, our benevolent feelings and sense of greater connection are also embodied in better physical health. As our awareness expands we become ever-more consciously connected and progressively resonate with the wholeness of the Cosmos.

The *principle of reflection* essentially extends the rule of resonance, describing how the outer circumstances of our lives are reflected inwardly by our mental, emotional, and eventually physical states. When we're able to consciously reflect on such mirroring, we're then more able to recognize and amend any distortions such reflection embeds.

All physical processes incorporate inevitable change through the inexorable flow of time; nothing stays the same. This *principle of change* when applied to our human circumstances calls us to embrace and learn the lessons of our experiences. As the Buddha noted, some of the greatest suffering we undergo in life is when our attachment to something or someone causes us to cling to situations, often until the pain of such holding on is greater than the real or imagined hurt of letting go.

We've seen how causes and effects play out ubiquitously through space-time Again, when this universal rule is applied to ourselves, the principle of *choice and consequence* takes on a deeper perspective, encouraging us to empower ourselves by taking responsibility for all our decisions. While clearly there are challenging situations in our lives that our human selves didn't consciously choose, nonetheless, we always have a choice in how to respond. Perhaps no one explained this truism more fundamentally than psychologist Viktor Frankl. Interred with other family members in concentration camps of the Holocaust and the only one to survive them, he later wrote that in every circumstance, even the horrendous ones that he lived through, "Between stimulus and response, there is a space. In that space is our power to choose our response. In our response lies our growth and our freedom."

The *principle of conservation* when applied to consciousness emphasizes the overall ebb and flow, give and take of life and is an extension of the conservation of energy. As for ebbs, flows, gives, takes, and cycles of energy, where energies change their forms but are ultimately conserved, this principle correlates to the ancient concept of karma. Significantly, though, it broadens its often-narrow tit-for-tat interpretation to nonjudgmentally encompass the entirety of our co-creative experiences on all levels of our consciousness.

The last of the octave is that of the *principle of concession,* which encourages us to recognize the underlying meaning and purpose of our life experiences and circumstances, to acknowledge responsibility for our choices, accept the appropriateness of what flows from them, and to learn and evolve accordingly.

Such concession is expressed by *ho'oponopono,* the ancient Hawaiian practice of forgiveness and reconciliation traditionally used to resolve family and tribal conflicts. Through a process of individual and collective acknowledgment of fault and recognition of responsibility, its prayer, "I'm sorry. Please forgive me. Thank you. I love you" enables the resolution, healing, and release of intrapersonal and interpersonal conflict.

As awareness of the unity of consciousness grows, the realization by the prayer of ho'oponopono that at some level of consciousness, each of us is a co-creator of the entirety of our reality and so ultimately responsible for its outcome, thus expands to encompass our entire Universe.

ME TO WE

Instead of viewing organisms as separate from and responding to a passive environment, evolutionary processes are, as we've seen, essentially co-creative, where the overall environment is in dynamic discourse with its biological expressions. Such a transformation in perception also substantially modifies the hitherto general consensus by biologists that the evolutionary fitness of organisms with their environments progresses solely through modification of the DNA genome, takes place only at individual levels, and is driven only by competition for scarce resources.

Increasingly, studies of epigenetic adaptations are showing that gene expression is modified by lifestyle and environmental factors even though the underlying DNA doesn't change.

As we'll go on to explore, in the past few years too, a growing number of researchers, and especially those working across disciplines, have proposed multilevel approaches to understanding evolution. These include not only co-creative adaptations between organisms and their environments but fitness selections also at kin and group levels and engendered by cooperative socialization.

Let's take a look at the burgeoning research on epigenetics first. Still in its early stages in terms of human biology, such investigations are nonetheless revealing that epigenetic traits not only have long-term effects in a living person but are able to be inherited by their children. Furthermore, if the studies of epigenetic factors in mice are replicated in people, such features may even persist through future generations.

In 2013 medical doctor Kerry Ressler and neurobiologist Brian Dias at Emory University caused male mice to fear the sweet smell of *acetophenome,* commonly found in fruit-tree blossoms, by making them associate it with electrical shocks. Not only did the shocked mice tremble when exposed to its scent, but their offspring did so too—even though they'd never previously encountered it nor suffered the shock treatment of their fathers. The fear even persisted to a third generation its memory informationally imprinted within their genetic expression.[4]

In two studies, psychiatrist Rachel Yehuda and her team at Mount Sinai hospital in New York have revealed their findings of such epigenetically transferred stress in human beings. Their first analysis, reported in 2005,[5] investigated the instances of post-traumatic stress disorder (PTSD) in pregnant women following the 9/11 attacks and whether such trauma epigenetically affected their children. Using their levels of the stress hormone cortisol as a measure, the researchers took samples from thirty-eight women who at the time of the attacks had been either at or near the World Trade Center. Those of the group who went on to suffer PTSD exhibited much lower cortisol levels than others who were similarly exposed but didn't develop PTSD. A year later when researchers took samples from their babies, the cortisol levels of the infants correlated with those of their mothers.

The second study, reported in 2015,[6] looked at the stress-related disorders of children of thirty-two Holocaust survivors and compared them to the children of Jewish families who had lived safely beyond Europe during World War II. The team investigated an epigenetic "tag" (a chemical attachment to DNA that switches genes on and off) specifically associated with the regulation of stress hormones. They discovered such a tag in both the Holocaust survivors and their children. Finding no such link with the children of the comparative group and carefully excluding the possibility that the survivors' children had undergone trauma themselves, they were able to show that such effects are indeed inherited and not just transferred by social influences or stressful events.

In contrast to the hitherto consensual view of biologists that evolution only progresses at the level of individuals and through innately selfish and conflictual behavior, a growing number of researchers are arguing instead for a more inclusive multilevel approach that recognizes the evolutionary effects of group cooperation and altruism.

In social species of insects such as ants and bees, their collective existence has evolved to depend on the mutual support driven by kin selection. The developing perspective, though, broadens and deepens such social collaboration in mammals. In melding individual selfishness with group altruism it's recognizing the evolutionary advantages conferred by such practical support and cultural relationships.

The realization that in higher animals and especially humans *emotional* bonds of kinship and within progressively wider groups are also vital to evolutionary progress and often a more powerful cause of altruistic behavior than any mundane considerations has, as yet, been hardly considered.

Nonetheless, such a so-called gene-culture coevolution approach is challenging the previous consensus that all evolutionary processes progress through conflict and solely through individual adaptation. As argued by anthropologist Robert Boyd and biologist Peter Richerson and others, such enriched evolutionary factors equip humans to rapidly develop specific adaptations, and far more quickly than genetic mutation alone.[7] Boyd and Richerson also maintain that the social development of collaboration in hominids, around a million years or so ago, evolved as a co-creative response to a period of rapid climate change and provided a behavioral template for future cooperative conducts.

Investigations by a number of researchers, such as economists Samuel Bowles and Herbert Gintis, who in 2003 published a statistical study, reveal that societies that embed such pro-social relationships have higher survival rates than those that don't.[8]

Recent analyses by social scientists Nicholas Christakis and James Fowler have also shown just how influential social networks are in changing the behaviors, beliefs, and even physical health of people who are three degrees of separation apart and who may have never met.[9] In such networks, ripples, ebbs, and flows of influence play out through the tenuous links of friends of friends of friends.

From our collective physical, mental, and emotional health to economic phenomena and the spread of innovation, the specific patterns of informational interactions within networks confer different properties upon the people enmeshed within them. These intra-connections tie the network into a form of "superorganism" whose whole is greater than the sum of its parts.

We've previously touched on the concept of small-world networks. The intermeshing of social networks, ubiquitous in human societies, innately embeds the "me" within the wider "we." With the Internet and distributed technologies that connect us in global networks we're now more collaboratively influenced than ever before. The benefits of our connected lives, Christakis maintains, outweigh the costs. As he says, "It is the spread of the good things that vindicates the whole reason we live our lives in networks."

With virtually all of us being intricately incorporated in our societies and networks, the findings of social epidemiologists Richard Wilkinson and Kate Pickett as described in their book *The Spirit Level,* first published in 2009, are particularly noteworthy.[10] Demonstrating, through a raft of socioeconomic, health-related, educational attainment, and criminal behavior data from twenty-three developed countries, they have shown that higher levels of inequality within a society engender greater levels of dysfunctionality—for everyone.

While equality of opportunity doesn't lead to equality of outcomes, owing to a vast array of factors, inbuilt inequalities drive additional inequality, where the playing field is not flat, but like Piff's game of monopoly, is rigged and where the outcome is inevitable. Unchecked, money buys power and influence, which then leads to greater power and influence and increasing inequality and all the consequential social ailments it causes.

As illustrated in a growing number of social studies over the past few years, such as those quoted by Wilkinson and Pickett, high levels of inequality reduce social mobility, societal coherence and trust, life expectancy, educational performance, and physical, mental, and emotional health, not just for the "99 percent," but for the "1 percent" too. We literally are the 100 percent.

14
Conscious Evolution

**The evolution of evolution:
from unconscious to conscious choice . . . defined
by futurist Barbara Marx Hubbard**

> *We shall make it our mission to design, communicate, and implement a more spiritual and harmonious civilization—a civilization that enables humankind to realize its inherent potential and advance to the next stage of its material, spiritual, and cultural evolution.*
>
> Fuji Declaration, 2014

This is a momentous time. The evolutionary journey of our perfect Universe has brought the self-awareness of the human race to a point where we are able to *consciously* evolve, and to do so in a timescale that may span mere generations rather than epochs.

Our choices now are crucial: not only to enable us to progress to the next level of perception but also to actually survive. Our hitherto limited perception has brought us and our planetary home to the edge of breakdown. In this global emergency, only the emergence of greater awareness will enable us to break through.

As indigenous elders around the world counsel us: "The choice is ours."

So what will it be?

OUR CHOICE

We are at a vital threshold in human history. While our fragmented perspective of the world has brought us and our planet to the edge of catastrophic breakdown, the newly emerging holistic perception of the cosmic hologram and the essential unity of consciousness offer us the choice for a transformational breakthrough.

The growing evidence goes far beyond the intellectual acknowledgment that everything in our perfect Universe is connected; the revelation of the cosmic hologram and its unification of reality literally in-forms the potential for our lived experience of its all-encompassing perspective. While retaining the uniqueness of our personal, microcosmic expressions of consciousness, this wholeworld-view embraces the meso-cosm of our collective human experience and the macrocosm of our entire Universe, existing as a finite expression of the infinity and eternity of cosmic mind.

When fully realized such a view of the world does away with the conflictual interactions of duality perception, empowers mitigation of selfishness, and enhances cooperation and altruism, not only with each other but with all life as well.

Crucially, comprehension of the unity of consciousness doesn't imply homogeneity. Instead it gives greater meaning to our personal sense of self while celebrating the diversity of our collective human expression, allowing us to perceive the profound purpose of all lives. When our awareness expands to encompass such wholeness, we resonate with, we attune to, we align with the harmony of our Universe's being. Our fears, emanating from the illusion of separation, are healed by the love that is our true foundation.

As we individually and collectively embrace this transformative perspective, our choices and behaviors too may be transformed and we are able to become conscious co-creators of our re-membered realities and the compassionate custodians of our planetary home. This is our personal and collective future—if we choose it.

INTO-GREAT

While spiritual traditions and mystics for millennia have explored other states of consciousness than our everyday waking awareness, only in the past half century or so have such altered and expanded perceptions been scientifically studied. Early investigators, including psychotherapist C. G. Jung, recognized that their case studies involving multidimensional awareness beyond that of the human ego were akin to Eastern mystical and indigenous shamanic practices and experiences. Their explorations of such transpersonal states have since been substantially extended by other pioneer investigators, notably psychologist Stanislav Grof.

In more than fifty years of research, Grof has catalogued the often independently verifiable experiences and related information of what he terms holotropic states, where numerous multidimensional and progressively archetypal and cosmic levels of consciousness are encountered.

In a series of books, including *When the Impossible Happens: Adventures in Non-Ordinary Realities,* Grof reports on copious profound experiences and communications with the consciousness and intelligence of many levels and realms of existence.[1] These include experiences of historic circumstances and events as well as experiences with past lives and deceased people; shamanic encounters and associative connections with other biological life-forms, including on microcosmic and collective levels; meetings with discarnate spirit guides and with elemental, devic, extraterrestrial, and angelic beings and associations with "mythic" entities. Experiential reports also describe the even more expansive realm of archetypes and cosmic principles that manifest in universal or culturally specific forms and, ultimately, the formless potentiality of the cosmic plenum itself.

Such multidimensional realms of consciousness have been called by many names throughout history; recently Ervin Laszlo has called these realms the *akashic* domain, taking his inspiration from the Hindu concept of *akasha* that from ancient times has described such essential transcendent realities.

There are three key findings of the ever more extensive research into such transpersonal phenomena.

First, rather like supernormal attributes, such non-ordinary experiences of multidimensional realities are able to access accurate and verifiable information at nonlocal levels.

Second, rather like many near-death reports, they open the portal to the realization that the human psyche continues after the demise of the physical body, so freeing death of its fear.

Third, by expanding awareness beyond its egoic restraints, they reveal the cosmic richness of consciousness playing out on myriad levels and realms of being.

While the hitherto mainstream science of dualism and materiality provides neither framework nor acceptance of such transcendence, the emerging understanding of the cosmic hologram enables multidimensional and nonlocal abilities, phenomena, and experiences to be encompassed within a comprehensive and comprehensible wholeworld-view.

In addition, as many researchers such as Grof, Laszlo, and philosopher Jean Houston have appreciated, not only the awareness of such expanded realms of reality but also communication with and learning from the intelligences that inhabit numerous levels of multidimensional existence offer a profoundly benevolent perception of and mentoring for the next step in our evolutionary progress.

ORDINARY EXTRAORDINARINESS

Living a transpersonal life, while grounding our multidimensional experiences within our physical human experience, is sometimes tricky, as I know from personal experience over nearly sixty years.

I first directly encountered the cosmic hologram of Indra's net and the realities of multidimensional consciousness when I was four years old. One day as I hovered between sleep and wakefulness, I had a vision that was as real as my bedroom in the home in the north of England that I shared with my coal-miner dad, housewife mum, grandmother, and not yet annoying younger brother.

In it, I seemed to be at the center of a vast interconnected and pulsing web of rainbow light, which shimmered in geometrical shapes that repeated and mirrored each other from the smallest to the largest scales I had a sense of and as far as I could see. Instead of merely being fixed, they changed from moment to moment, and I became aware that there were living forms of light made up from their patterns.

Since that first revelation, numerous psychic perceptions, altered states of consciousness, out-of-body experiences, and the validation of what they've taught me and the insights I've gained from them have, over my lifetime, convinced me of such realities far beyond the physical level of existence and of the interconnected unity of an intelligent Cosmos.

It was seeking to understand the wider nature of reality that these many mystical occurrences have profoundly revealed that first impelled me to ask not only *how* but also *why* the Cosmos is as it is. They've progressively led me to see reality as the consciousness of an infinite cosmic mind being dynamically co-created and experienced on multidimensional levels of existence. And over many years these experiences have also resulted in my perceiving the physical realm as being holographically and holarchically realized at all scales of existence—from the ultimate unity of the whole world and the infinite intelligence of cosmic mind.

The entirety of my experiential understanding is encapsulated in the emerging scientifically based concept of the cosmic hologram and the rapidly increasing and wide-ranging evidence for it as set out in this book.

This journey of discovery and re-membering is still ongoing. The only criteria that I personally adhere to, inspired by the wisest pioneering scientists and the most courageous mystics, is to keep an open mind and an open heart and be willing to follow the evidence, wherever it may lead.

✻

While the journey of discovery and re-membering goes on, the way-showers for our progress onward toward understanding, experiencing, and ultimately embodying the unity awareness of the wholeworld-view (visit www.wholeworld-view.org) are coming ever more clearly into view.

- In-formation *is* reality
- It *is* bit
- Mind *is* matter
- Matter *is* mind

And mind and consciousness aren't something we have; they're what we and the whole world *are*.

Acknowledgments

I could not have written this book without the incredible dedication of numerous scientists, philosophers, and spiritual seekers in undertaking their own journeys of discovery into the deeper nature of reality.

My gratitude to all those pioneering scientific thinkers who have been willing to follow the evidence wherever it has led them, and here to especially thank Isaac Newton, Ludwig Boltzmann, James Clerk Maxwell, Amelie Noether, Max Planck, Albert Einstein, Alan Turing, Dennis Gabor, Claude Shannon, David Bohm, John Archibald Wheeler, Benoit Mandelbrot, and writer Michael Talbot.

I offer my gratitude too to those philosophers and spiritual seekers from all traditions who, throughout the ages, have explored deeply into the cosmic hologram of reality and shared their perceptions of universal spiritual experiences.

My thanks to my friends and colleagues of the Evolutionary Leaders circle who offer amazing support, wise counsel, and joyous friendship and to the many others of an ever-growing community around the world, who serve the emergence of conscious evolution.

My thanks to my literary agent, Susan Mears, for her expertise and enduring commitment to raising our collective awareness.

My gratitude to all those at Inner Traditions who have brought such proficiency, dedication, and care to this book and to sharing its message, especially acquisitions editor Jon Graham, my editors Jennie Marx and

Cannon Labrie, sales and marketing director John Hays, and publicist Blythe Bates.

My appreciation to Ervin Laszlo, not only for kindly writing the foreword to the book but also for his and his beloved Carita's continuing friendship and support.

My thanks to my friend, colleague, and fellow traveler Gil Agnew, who now walks beside me as we aim to share the seed-point message of *The Cosmic Hologram* and the two further books in the Transformation trilogy of the wholeworld-view of unified reality.

During my life-time's explorations and experiences of the wholeworld I've been enormously privileged to meet, learn from, and be guided by wisdomkeepers of many traditions—both incarnate and discarnate. While too many to name, my gratitude to you all remains in my heart.

Every day, I give wholehearted thanks for my treasured husband and soul mate, Tony, who shares and supports all my life's adventures with good humor, patience, and love. He makes me laugh, shelters my spirit, hugs me when the world feels heavy, and always uplifts and inspires me with his amazing example.

Finally, I offer gratitude almost beyond words to Thoth, who appeared as a discarnate light in my room when I was four years old and has been with me ever since: my mentor, guide, and dear friend.

✷

Starting in May 2017, Jude will be teaching an online course relating to *The Cosmic Hologram* at Ubiquity University.

Notes

PREFACE. INDRA'S NET

1. G. 't Hooft, "Canonical Quantization of Gravitating Point Particles in 2+1 Dimensions," *Classical and Quantum Gravity* 10, no. 8 (1993): 1653. arXiv:gr-qc/9305008.

CHAPTER I. IN-FORMATION

1. J. A. Frieman, M. S. Turner, and D. Huterer, "Dark Energy and the Accelerating Universe," *Annual Review of Astronomy and Astrophysics* 46, no. 1 (2008): 385–432. arXiv:0803.0982. doi:10.1146/annurev.astro.46.060407.145243.
2. R. Landauer, "Information Is Physical," *Physics Today* 44 (1991): 23–29.
3. L. Szilard, "On the Decrease of Entropy in a Thermodynamic System by the Intervention of Intelligent Beings" (trans.), *Zeitschrift für Physik* 53 (1929): 840–56. www.sns.ias.edu/~tlusty/courses/InfoInBioParis/Papers/Szilard1929.pdf.
4. A. Bérut, A. Arakelyan, A. Petrosyan, S. Ciliberto, R. Dillenschneider, and E. Lutz, "Experimental Verification of Landauer's Principle Linking Information and Thermodynamics, *Nature* 483 (2012): 187–89.
5. A. Peruzzo, P. Shadbolt, N. Brunner, S. Popescu, and J. L. O'Brien, "A Quantum Delayed Choice Experiment," *Science* 338 (2012): 634–37. http://arxiv.org/pdf/1205.4926.pdf.
6. S. S. Afshar, "Waving Copenhagen Good-bye: Were the Founders of Quantum Mechanics Wrong?" Harvard seminar announcement (2004).

7. R. Menzel, D. Puhlmann, A. Heuer, and W. P. Schleich, "Wave-Particle Dualism and Complementarity Unraveled by a Different Mode," Proceedings of the National Academy of Sciences of the United States of America 109, no. 24 (2012): 9314–319. www.pnas.org/content/109/24/9314.abstract.
8. E. Bolduc, J. Leach, F. M. Miatto, G. Leuchs, and R. W. Boyd, "Fair Sampling Perspective on an Apparent Violation of Duality," Proceedings of the National Academy of Sciences of the United States of America (2014).
9. H. Everett III, "Relative State Formulation of Quantum Mechanics," *Reviews of Modern Physics* (1957).
10. https://en.wikipedia.org/wiki/Dennis_Gabor.

CHAPTER 2. INSTRUCTIONS

1. L. M. Krauss, *A Universe From Nothing* (New York: Simon & Schuster, 2012).
2. C. Blake et al., "The WiggleZ Dark Energy Survey: Measuring the Cosmic Expansion History Using the Alcock-Paczynski Test and Distant Supernovae," *Astronomy and Geophysics* 49, no. 5 (2011): 5.19–5.24. https://arxiv.org/pdf/1108.2637.pdf.
3. P. A. Milne, R. J. Foley, P. J. Brown, and G. Narayan, "The Changing Fractions of Type Ia Supernova NUV—Optical Subclasses with Redshift," *Astrophysical Journal* 803, no. 20 (2015). doi:10.1088/0004-637X/803/1/20.
4. A. S. Eddington, *The Nature of the Physical World* (New York: The MacMillan Company, 1915), 74.
5. S. Toyabe, T. Sagawa, M. Ueda, E. Muneyuki, and M. Sano, "Information Heat Engine: Converting Information to Energy by Feedback Control," *Nature Physics* 6 (2010): 988–92. http://arxiv.org/pdf/1009.5287.pdf.
6. J. D. Bekenstein, "Universal Upper Bound on the Entropy-to-Energy Ratio for Bounded Systems," *Physical Review D* 23, no. 2 (January 15, 1981): 287–98. doi:10.1103/PhysRevD.23.287.

CHAPTER 3. CONDITIONS

1. J. M. Maldacena, "The Large N Limit of Superconformal Field Theories and Supergravity," *Advanced Theoretical Math and Physics* 2 (1998): 231–52. https://arxiv.org/abs/hep-th/9711200.
2. A. Aspect, P. Grangier, and G. Roger, "Experimental Realization of

Einstein-Podolsky-Rosen-Bohm Gedankenexperiment: A New Violation of Bell's Inequalities," *Physical Review Letters* 49, no. 2 (1982): 91–94. doi:10.1103/PhysRevLett.49.91.

3. F. Bussières, C. Clausen, A. Tiranov, B. Korzh, V. B. Verma, S. W. Nam, F. Marsili, A. Ferrier, P. Goldner, H. Herrmann, C. Silberhorn, W. Sohler, M. Afzelius, and N. Gisin, "Quantum Teleportation from a Telecom-Wavelength Photon to a Solid-State Quantum Memory," *Nature Photonics* 8 (2014): 775–78. http://arxiv.org/pdf/1401.6958.pdf.
4. K. C. Lee, M. R. Sprague, B. J. Sussman, J. Nunn, N. K. Langford, X. M. Jin, T. Champion, P. Michelberger, K. F. Reim, D. England, D. Jaksch, and I. A. Walmsley, "Entangling Macroscopic Diamonds at Room Temperature," *Science* 334, no. 6060 (2011): 1253–256. doi:10.1126/science.1211914.

CHAPTER 4. INGREDIENTS

1. D. Hutsemékers, L. Braibant, V. Pelgrims, and D. Sluse, "Spooky Alignment of Quasars Across Billions of Light-Years," *European Southern Observatory* (November 19, 2014). www.eso.org/public/news/eso1438.
2. K. N. Abazajian, N. Canac, S. Horiuchi, M. Kaplinghat, and A. Kwa, "Discovery of a New Galactic Center Excess Consistent with Upscattered Starlight," *Journal of Cosmology and Astroparticle Physics* 7 (2015). http://arxiv.org/abs/1410.6168 (submitted 22 Oct. 2014; last revised 10 Jul. 2015).
3. A. Bogdan and A. Goulding, "Dark Matter Guides Growth of Supermassive Black Holes," Harvard-Smithsonian Center for Astrophysics release 2015-07 (February 18, 2015).
4. V. Salvatelli, N. Said, M. Bruni, A. Melchiorri, and D. Wands, "Indications of a Late-Time Interaction in the Dark Sector," *Physical Review Letters* 113 (October 30, 2014).
5. R. L. Jaffe, "The Casimir Effect and the Quantum Vacuum," *Physical Review* D72 (2005). https://arxiv.org/abs/hep-th/0503158 (submitted Mar. 21, 2005).
6. S. J. Brodsky, C. D. Roberts, R. Shrock, and P. C. Tandy, "New Perspectives on the Quark Condensate," *Physical Review C* 82, 022201(R). www.slac.stanford.edu/th/lectures/Stanford_DarkEnergy_B_Dec2010.pdf.
7. Laser Interferometer Gravitational-Wave Observatory, "Gravitational Waves Detected 100 Years After Einstein's Prediction," February 11, 2016, www.ligo.caltech.edu/news/ligo20160211.
8. T. Jacobson, "Thermodynamics of Spacetime: The Einstein Equation of

State," *Physical Review Letters* 75 (1995): 1260–263. http://arxiv.org/pdf/gr-qc/9504004.pdf.

9. E. P. Verlinde, "On the Origin of Gravity and the Laws of Newton," *Journal of High Energy Physics* (2011). https://arxiv.org/abs/1001.0785.
10. T. Wang, "Modified Entropic Gravity Revisited," *High Energy Physics Theory.* https://arxiv.org/abs/1211.5722 (submitted November 25, 2012).
11. R. Loll, "What You Always Wanted to Know about CDT, but Did Not Have Time to Read about in Our Papers," mp4 seminar. http://pirsa.org/14040086 (November 18, 2005).
12. P. Hořava, "Quantum Gravity at a Lifshitz Point," *Physical Review D* 79, no. 8 (2009). arXiv:0901.3775.
13. P. G. O. Freund, "Emergent Gauge Fields," *High Energy Physics.* http://arxiv.org/abs/1008.4147 (submitted August 24, 2010).
14. Z. Chang, M-H. Li, and X. Li, "Unification of Dark Matter and Dark Energy in a Modified Entropic Force Model," *Commun. Theoretical Physics* 56 (2011): 184–92. http://arxiv.org/abs/1009.1506.

CHAPTER 5. RECIPE

1. SLAC, "BaBar Experiment Confirms Time Asymmetry." www6.slac.stanford.edu/news/2012-11-19-babar-trv.aspx.
2. M. Rees, *Just Six Numbers: The Deep Forces That Shape the Universe* (New York: Basic Books, 2001).
3. M. P. Mueller and L. Masanes, "Three-Dimensionality of Space and the Quantum Bit: An Information-Theoretic Approach," *New Journal of Physics* 15 (2013). http://arxiv.org/abs/1206.0630.
4. B. Dakic, T. Paterek, and C. Brukner, "Density Cubes and Higher-Order Interference Theories," *New Journal of Physics* 16 (2013). http://arxiv.org/pdf/1308.2822v2.pdf.

CHAPTER 6. CONTAINER

1. Y. Yuval, M. Eitan, Z. Iluz, Y. Hanein, A. Boag, and J. Scheuer, "Highly Efficient and Broadband Wide-Angle Holography Using Patch-Dipole Nanoantenna Reflectarrays," *Nano Letters* 14, no. 5 (2014): 2485. doi:10.1021/nl5001696.
2. X. Xu, X. Liang, Y. Pan, R. Zheng, and Z. A. Lum, "Spatiotemporal

Multiplexing and Streaming of Hologram Data for Full-Color Holographic Video Display," *Optical Review* 21 (February 2015): 220–25.

3. B. Long, S. A. Seah, T. Carter, and S. Subramanian, "Rendering Volumetric Haptic Shapes in Mid-Air Using Ultrasound," *ACM Transactions on Graphics* (November 2014). http://dx.doi.org/10.1145/2661229.2661257.
4. Fermilab, "The Holometer: A Fermilab Experiement," YouTube video (posted December 16, 2014). www.youtube.com/watch?v=8HqEaPKZ7fs.
5. A. S. Chou, R. Gustafson, C. Hogan, B. Kamai, O. Kwon, R. Lanza, L. McCuller, S. S. Meyer, J. Richardson, C. Stoughton, R. Tomlin, S. Waldman, and R. Weiss, "Search for Space-Time Correlations from the Planck Scale with the Fermilab Holometer," *Fermilab* (December 2015). https://arxiv.org/pdf/1512.01216.pdf.
6. Planck Collaboration, "Planck 2015 results. XIII. Cosmological Parameters." *Astronomy and Astrophysics* (February 2015). https://arxiv.org/abs/1502.01589.
7. Y. B. Zeldovich and A. Starobinski, "Quantum Creation of a Universe with Nontrivial Topology," *Soviet Astronomy Letters* 10, no. 135 (1984).
8. M. Tegmark, A. de Oliveira-Costa, and A. Hamilton, "A High Resolution Foreground Cleaned CMB Map from WMAP," *Physical Review* D68 (2003). http://arxiv.org/abs/astro-ph/0302496.
9. M. M. Caldarelli, J. Camps, B. Goutéraux, and K. Skenderis, "AdS/Ricci-Flat Correspondence and the Gregory-Laflamme Instability," *Physical Review* D67 (March 2013). http://eprints.soton.ac.uk/391645.
10. R. Aurich, H. S. Janzer, S. Lusti and F. Steiner, "Do We Live in a Small Universe?" *Classical and Quantum Gravity* 25 (2008). http://arxiv.org/abs/0708.1420.

CHAPTER 8. UNIVERSAL PATTERNS

1. D. L. Turcotte, "Fractals in Geology: What Are They and What Are They Good For?" *GSA Today* (1991). www.geosociety.org/gsatoday/archive/1/1/pdf/i1052-5173-1-1-sci.pdf.
2. "Fractal Patterns Spotted in the Quantum Realm," Physics World online (February 9, 2010). http://physicsworld.com/cws/article/news/2010/feb/09/fractal-patterns-spotted-in-the-quantum-realm.
3. B. Hunt, J. D. Sanchez-Yamagishi, A. F. Young, M. Yanlowitz, B. J. LeRoy, K. Watanabe, T. Taniguchi, P. Moon, M. Koshino, P. Jarillo-Herrero, and

R. C. Ashoori, "Massive Dirac Fermions and Hofstadter Butterfly in a van der Waals Heterostructure," *Science* 340, no. 6139 (June 2013): 1427–30. doi:10.1126/science.1237240.

4. M. Fratini, N. Poccia, A. Ricci, G. Campi, M. Burghammer, G. Aeppli, and A. Bianconi, "Scale-Free Structural Organization of Oxygen Interstitials in La_2CuO_{4+y}," *Nature* 466 (August 2010): 841–44. www.nature.com/articles/nature09260.epdf.
5. Warwick University, "Astrophysicists Find Fractal Image of Sun's 'Storm Season' Imprinted on Solar Wind" (2014). www2.warwick.ac.uk/newsandevents/pressreleases/astrophysicists_find_fractal.
6. J. Li and M. Ostoja-Starzewski, "Saturn's Rings Are Fractal," (June 2012). https://arxiv.org/abs/1207.0155.
7. International Centre for Radio Astronomy Research, "WiggleZ Confirms the Big Picture of the Universe" (2012). www.icrar.org/news/news_items/media-releases/wigglez-confirms-the-big-picture-of-the-universe.
8. L. McClelland, T. Simkin, M. Summers, E. Nielsen, T. C. Stein, eds. *Global Volcanism 1975–1985* (Englewood Cliffs, NJ: Prentice Hall, and Washington, D.C.: American Geophysical Union, 1989).
9. J. F. Lindner, V. Kohar, B. Kia, M. Hippke, J. G. Learned, and W. L. Ditto, "Strange Nonchaotic Stars," *Physical Review Letters* 114 (2015): 1–5.
10. L. P. Kadanoff, "The Droplet Model and Scaling," in *Critical Phenomena, Proceedings of the Int. School of Physics,* edited by M. S. Green (New York: Academic Press, 1971), 118–22.
11. P. Bak, C. Tang, and K. Wiesenfeld, "Self-organized Criticality: An Explanation of the 1/*f* Noise," *Physical Review Letters* 59 (July 1987): 381–84.
12. E. Lorenz, "Predictability; Does the Flap of a Butterfly's Wings in Brazil Set Off a Tornado in Texas?" American Association for the Advancement of Science 139th Meeting (1972). http://eaps4.mit.edu/research/Lorenz/Butterfly_1972.pdf.

CHAPTER 9. IN-FORMED DESIGN *FOR* EVOLUTION

1. J. P. Crutchfield and D. P. Feldman," "Regularities Unseen, Randomness Observed: The Entropy Convergence Hierarchy," *Chaos* 15 (2003): 25–54.
2. D. P. Feldman, C. S. McTague, and J. P. Crutchfield, "The Organization of Intrinsic Computation: Complexity-Entropy Diagrams and the Diversity of

Natural Information Processing," *Chaos* 18 (2008). doi:10.1063/1.2991106. Also available at: arXiv:0806.4789v1.

3. B. Skyrms, "Signals, Evolution and the Explanatory Power of Transient Information," *Philosophy of Science* 69, no. 3 (2002): 407–28.
4. J. J. Johnson, A. Tolk, and A. Sousa-Poza, "A Theory of Emergence and Entropy in Systems of Systems," *Procedia Computer Science* 20 (2013): 283–89.
5. https://en.wikipedia.org/wiki/Evaporating_gaseous_globule.
6. Harvard-Smithsonian Center for Astrophysics, "Magnetic Fields Play a Larger Role in Star Formation than Previously Thought," news release, September 9, 2009. www.cfa.harvard.edu/news/2009-20. Relating to: H-b. Li, D. Dowell, A. Goodman, R. Hildebrand, and G. Novak, "Anchoring Magnetic Field in Turbulent Molecular Clouds," *The Astrophysical Journal* 704, no. 2 (2009). http://arxiv.org/abs/0908.1549.
7. Max Planck Institute for Radio Astronomy, "Interstellar Molecules are Branching Out," news release, September 25, 2014. www.mpifr-bonn.mpg .de/pressreleases/2014/10.
8. C. L. Ilsedore, E. A. Bergin, C. L. O'D. Alexander, F. Du, D. Graninger, K. J. Oberg, and T. J. Harries, "The Ancient Heritage of Water Ice in the Solar System," *Science* 345 (2014). doi:10.1126/science.1258055.
9. T. B. Mahajan, J. E. Elsila, D. W. Deamer, and R. N. Zare, "Formation of Carbon-Carbon Bonds in the Photochemical Alkylation of Polycyclic Aromatic Hydrocarbons," *Origins of Life and Evolution of Biospheres* 33 (2002): 17. web.stanford.edu/group/Zarelab/publinks/zarepub677.pdf.
10. P. Michael, M. P. Callahan, K. E. Smith, H. J. Cleaves II, J. Ruzicka, J. C. Stern, D. P. Glavin, C. H. House, and J. P. Dworkin, "Carbonaceous Meteorites Contain a Wide Range of Extraterrestrial Nucleobases," *Proceedings of the National Academy of Sciences* 108, no. 34 (2011): 13995–998. http://www.pnas.org/content/108/34/13995.short.
11. J. K. Jorgensen, C. Favre, S. E. Bisschop, T. L. Bourke, E. F. van Dishoek, and M. Schmalzl, "Detection of the Simplest Sugar, Glycolaldehyde, in a Solar-Type Protostar with ALMA," *Astrophysics Journal Letters* 757 (2012). www.eso.org/public/archives/releases/sciencepapers/eso1234/eso1234a .pdf.
12. R. Malhotra, "Orbital Resonances and Chaos in the Solar System," in *Solar System Formation and Evolution, ASP Conference Series,* vol. 149, edited by D. Lazzaro et al. (1998).

13. K. Batygin and G. Laughlin, "Jupiter's Decisive Role in the Inner Solar System's Early Evolution," *Proceedings of the National Academy of Sciences* 112, no. 14 (2015): 4214–4217. www.pnas.org/content/112/14/4214.abstract.
14. J. Hecht, "Saturn's Calming Nature Keeps Earth Friendly for Life," *New Scientist* November 21, 2014. www.newscientist.com/article/dn26601-saturns-calming-nature-keeps-earth-friendly-to-life. Based on E. Pilat-Lohinger, "The Role of Dynamics on the Habitability of an Earth-like Planet," *International Journal of Astrobiology* 14, no. 2 (special issue, 2015): 145–52.
15. M. Landeau, P. Olsen, R. Degeun, and B. H. Hirsch, "Core Merging and Stratification Following Giant Impact," *Nature Geoscience,* published online September 12, 2016. www.nature.com/ngeo/journal/vaop/ncurrent/full/ngeo2808.html.
16. C. W. Carter Jr. and R. Wolfenden, "tRNA Acceptor Stem and Anticodon Bases Form Independent Codes Related to Protein Folding," *Proceedings of the National Academy of Sciences* 112, no. 24 (2015): 7489–494. www.pnas.org/content/112/24/7489.full.pdf; R. Wolfenden, C. A. Lewis Jr., Y. Yuan, and C. W. Carter Jr., "Temperature dependence of amino acid hydrophobicities," *Proceedings of the National Academy of Sciences* (2015). doi:10.1073/pnas.1507565112.
17. B. H. Patel, C. P. Percivalle, D. J. Ritson, C. D. Duffy, and J. D. Sutherland, "Common Origins of RNA, Protein and Lipid Precursors in a Cyanosulfidic Protometabolism," *Nature Chemistry* 7 (2015): 301–7. www.nature.com/nchem/journal/v7/n4/full/nchem.2202.html.
18. T. Shomrat and M. Levin, "An Automated Training Paradigm Reveals Long-Term Memory in Planaria and Its Persistence through Head Regeneration," *Journal of Experimental Biology* 216, no. 20 (2013): 3799–810. doi:10.1242/jeb.087809.
19. K. Burton, "NASA Scientists Find Clues that Life Began in Deep Space," NASA news release January 26, 2001. www.nasa.gov/centers/ames/news/releases/2001/01_06AR.html.
20. B. H. Lipton, *The Biology of Belief: Unleashing the Power of Consciousness, Matter and Miracles* (Carlsbad, Calif.: Hay House, 2011).
21. https://en.wikipedia.org/wiki/Stuart_Kauffman.
22. R. H. Thompson and L. W. Swanson, "Hypothesis-Driven Structural Connectivity Analysis Supports Network over Hierarchical Model of Brain

Architecture," *Proceedings of the National Academy of Sciences* 107, no. 34 (2010): 15235–239. www.ncbi.nlm.nih.gov/pubmed/20696892.

23. https://en.wikipedia.org/wiki/Milankovitch_cycles.

CHAPTER 10. HOLOGRAPHIC BEHAVIORS

1. W. Willinger and V. Paxson, "Where Mathematics Meets the Internet," *Notices of the American Mathematical Society* 45 (1998): 961–70.
2. R. Albert, H. Jeong, and A-L. Barabási, "The Diameter of the WWW," *Nature* 401 (1999): 130–31. arXiv:cond-mat/9907038.
3. M. Faloutsos, P. Faloutsos, and C. Faloutsos, *Power-Laws of the Internet*, Technical Report UCR-CS-99-01 (Riverside: University of California, 1999).
4. L. F. Richardson, "Variation of the Frequency of Fatal Quarrels with Magnitude," *Journal of the American Statistical Association* 43, no. 244 (1948): 523–46.
5. L. F. Richardson, "Statistics of Deadly Quarrels, 1809–1949," ICPSR5407. www.icpsr.umich.edu/icpsrweb/ICPSR/studies/5407 (pub. 1984).
6. Miami University, "Predicting Insurgent Attacks," news release July 14, 2011. www.miami.edu/index.php/news/releases/predicting_insurgent_attacks.
7. D. J. Watts and S. H. Strogatz, "Collective Dynamics of 'Small-World' Networks," *Nature* 393 (1998): 440–42. doi:10.1038/30918.
8. A-L Barabási and J. G. Oliveira, "Human Dynamics: Darwin and Einstein Communication Patterns," *Nature* 437 (2005). www.nature.com/nature/journal/v437/n7063/abs/4371251a.html.
9. Z. Dezsö, E. Almaas, A. Lukács, B. Rácz, I. Szakadát, and A-L Barabási, "Dynamics of Information Access on the Web," *Physical Review* 73 (2006).
10. D. Rybski, S. V. Buldyrev, S. Havlin, F. Lilijeros, and H. A. Makse, "Scaling Laws of Human Interaction Activity," *Proceedings of the National Academy of Sciences* 106, no. 31 (2009): 12640–645. www.pnas.org/content/106/31/12640.abstract.
11. C. Fan, J-L. Guo, and Y-L. Zha, "Fractal Analysis on Human Behaviors Dynamics," *Physica A* 391 (2012): 6617–625. http://arxiv.org/ftp/arxiv/papers/1012/1012.4088.pdf.
12. C. Song, Z. Qu, N. Blumm, and A-L. Barabási, "Limits of Predictability in Human Mobility," *Science* 327, no. 5968 (2010): 1018–21. doi:10.1126/science.1177170.

13. M. Sambridge, H. Tkalčić, and A. Jackson, "Benford's Law in the Natural Sciences," *Geophysical Research Letters* 37 (2010). doi:10.1029/2010GL044830.
14. J. Aron. "Mathematical Crime-fighter Helps Hunt for Alien Worlds," *New Scientist,* November 28, 2013. www.newscientist.com/article/dn24668-mathematical-crime-fighter-helps-hunt-for-alien-worlds.
15. X. Gabaix, "Zipf's Law for Cities: An Explanation," *The Quarterly Journal of Economics* 114, no. 3 (August 1999): 739–67. www.jstor.org/stable/2586883.
16. H. Lin and A. Loab, "Astrophysicists Prove that Cities on Earth Grow in the Same Way as Galaxies in Space," *MIT Technology Review,* January 16, 2015. www.technologyreview.com/s/534251/astrophysicists-prove-that-cities-on-earth-grow-in-the-same-way-as-galaxies-in-space.
17. W. Cheng, P. K. Law, H. C. Kwan, and R. S. S. Cheng, "Stimulation Therapies and the Relevance of Fractal Dynamics to the Treatment of Diseases," *Open Journal of Regenerative Medicine* 3, no. 4 (2014): 73–94. www.scirp.org/journal/PaperInformation.aspx?paperID=51401.
18. N. N. Taleb, *The Black Swan: The Impact of the Highly Improbable,* 2nd ed. (New York: Random House, 2010).

CHAPTER 11. *WHO* MAKES OUR PERFECT UNIVERSE?

1. D. J. Simons and D. L. Levin, "Failure to Detect Changes to People During a Real-World Interaction," *Psychonomic Bulletin & Review* 5 no. 4 (1998): 644–49.
2. W. J. Cromie, "Meditation Changes Temperature: Mind Controls Body in Extreme Experiments," *Harvard Gazette,* April 18, 2002.
3. E. Peper, V. S. Wilson, M. Kawakami, and M. Sata, "The Physiological Correlates of Body Piercing by a Yoga Master: Control of Pain and Bleeding," *Subtle Energies and Energy Medicine Journal* 14, no. 3 (2005): 223–37. https://biofeedbackhealth.files.wordpress.com/2011/01/final-piercing-7-15-05.pdf.
4. V. De Pascalis, "Psychophysiological Correlates of Hypnosis and Hypnotic Susceptibility," *International Journal of Clinical Experimental Hypnosis* 47, no. 2 (1999): 117–43.
5. G. H. Montgomery and I. Kirsch, "Mechanisms of Placebo Pain Reduction: An Empirical Investigation," *Psychological Science* 7 (1996): 174–76.

6. F. Benedetti, J. Durando, and S. Vighetti, "Nocebo and Placebo Modulation of Hypobaric Hypoxia Headache Involves the Cyclooxygenase-Prostaglandins Pathway," *Pain* 155, no. 5 (May 2014): 921–28. www.ncbi.nlm.nih.gov/pubmed/24462931.
7. S. Silberman, "Placebos Are Getting More Effective. Drugmakers Are Desperate to Know Why," *Wired,* August 24, 2009. www.wired.com/2009/08/ff-placebo-effect.
8. K. Paramaguru, "Has the Universe Stopped Producing New Stars?" *Time,* November 13, 2012. http://newsfeed.time.com/2012/11/13/has-the-universe-almost-stopped-producing-new-stars.
9. N. J Poplawski, "Cosmology with Torsion: An Alternative to Cosmic Inflation," *Physics Letters B* 694 (2010): 181–85. https://arxiv.org/abs/1007.0587.
10. M. J. Longo, "Detection of a Dipole in the Handedness of Spiral Galaxies with Redshifts $z \sim 0.04$," *Physics Letters B* 699, no. 4 (May 2011): 224–29.

CHAPTER 12. SUPERNORMAL

1. D. Ellerman, "A Common Fallacy in Quantum Mechanics: Why Delayed Choice Experiments Do NOT Imply Retrocausality" (2012). http://jamesowenweatherall.com/SCPPRG/EllermanDavid2012Man_QuantumEraser2.pdf. Published as "Why Delayed Choice Experiments Do NOT Imply Retrocausality," *Quantum Studies: Mathematics and Foundations* 2, no. 2 (2015): 183–99.
2. Colin Cooke, *An Introduction to Experimental Physics* (London: UCL Press, 1996), 5.
3. E. Laszlo and J. Currivan, *CosMos: A Co-Creator's Guide to the Whole World* (Carlsbad, Calif.: Hay House, 2008).
4. Benedict Carey, "A Princeton Lab on ESP Plans to Close Its Doors," *New York Times,* February 10, 2007. www.nytimes.com/2007/02/10/science/10princeton.html?_r=1.
5. L. Storm, P. E. Tressoldi, and L. Di Risio, "Meta-Analysis of Free-Response Studies, 1992–2008: Assessing the Noise-Reduction Model in Parapsychology," *Psychological Bulletin* 136, no. 4 (2010): 471–85.
6. L. Storm, P. E. Tressoldi, and L. Di Risio, "Meta-Analysis of ESP Studies, 1987–2010: Assessing the Success of the Forced-Choice Design in Parapsychology," *Journal of Parapsychology* 76, no. 2 (2012): 243–74.

7. L. Storm, P. E. Tressoldi, and J. Utts, "Testing the Storm et al. (2010) Meta-Analysis Using Bayesian and Frequentist Approaches: Reply to Rouder et al.," *Psychological Bulletin* 139, no. 1 (2013): 248–54.
8. P. E. Tressoldi, "Extraordinary Claims Require Extraordinary Evidence: The Case of Non-Local Perception, a Classical and Bayesian Review of Evidences," *Frontiers in Psychology* 2, no. 117 (June 2011). doi:10.3389/fpsyg.2011.00117.
9. J. Mossbridge, P. E. Tressoldi, and J. Utts, "Predictive Physiological Anticipation Preceding Seemingly Unpredictable Stimuli: A Meta-Analysis," *Frontiers in Psychology* 3, no. 390 (October 2012). doi:10.3389/fpsyg.2012.00390.
10. S. Schmidt, "Can We Help Just by Good Intentions? A Meta-Analysis of Experiments on Distant Intention Effects," *Journal of Alternative and Complementary Medicine* 18, no. 6 (2012): 529–33. doi:10.1089/acm.2011.0321.
11. Guy Lyon Playfair, *Twin Telepathy: The Psychic Connection* (3rd edition, Hove, UK: White Crow Books, 2012).
12. S. Saseendran, "Miracle Girl: Nandana Has Access to Mother's Memory," *Khaleej Times,* March 15, 2013. www.khaleejtimes.com/business/miracle-girl-nandana-has-access-to-mother-s-memory.
13. D. Radin, *Entangled Minds: Extrasensory Experiences in a Quantum Reality* (New York: Paraview Pocket Books/Simon & Schuster, 2006).
14. D. J. Bierman and H. S. Scholte, "Anomalous Anticipatory Brain Activation Preceding Exposure of Emotional and Neutral Pictures," University of Amsterdam (2002).
15. R. McCraty, M. Atkinson, and R. T. Bradley, "Electrophysiological Evidence of Intuition: Part 1. The Surprising Role of the Heart," *Journal of Alternative and Complementary Medicine* 10, no. 1 (2004): 133–43.
16. D. J. Bem, "Feeling the Future: Experimental Evidence for Anomalous Retroactive Influences on Cognition and Affect," *Journal of Personality and Social Psychology* 100, no. 3 (March 2011): 407–25. doi:10.1037/a0021524.
17. S. Rhine Feather and M. Schmicker, *The Gift: Extraordinary Experiences of Ordinary People* (New York: St. Martin's, 2006).
18. S. A. Schwartz, *Opening to the Infinite* (Buda, Tex: Nemoseen Media, 2007).
19. E. C. May, S. J. P. Spottiswoode, and L. V. Faith, "The Correlation of

the Gradient of Shannon Entropy and Anomalous Cognition," *Journal of Scientific Exploration* 14, no. 1 (2000): 53–72.

20. S. Parnia, K. Spearpoint , G. de Vos, P. Fenwick, D. Goldberg, J. Yang, J. Zhu, K. Baker, H. Killingback, P. McLean, M. Wood, A. M. Zafari, N. Dickert, R. Beisteiner, F. Sterz, M Berger, C. Warlow, S. Bullock, S. Lovett, R. M. McPara, S. Marti-Navarette, P. Cushing, P. Wills, K. Harris, J. Sutton, A. Walmsley, C. D. Deakin, P. Little, M. Farber, B. Greyson, and E. R. Schoenfeld, "AWARE-AWAreness During REsuscitation: A Prospective Study," *Resuscitation* (2014). www.ncbi.nlm.nih.gov/pubmed/25301715.
21. University of Southampton, "Results of World's Largest Near Death Experiences Study Published," October 7, 2014 press release. www.southampton.ac.uk/news/2014/10/07-worlds-largest-near-death-experiences-study.page.
22. P. van Lommel, R. van Wees, V. Meyers, and I. Elfferich, "Near-Death Experience in Survivors of Cardiac Arrest: A Prospective Study in the Netherlands," *Lancet* 358, no. 9298 (December 2001): 2039–45.
23. E. Cardeña, "A Call for an Open, Informed Study of All Forms of Consciousness," *Frontiers in Human Neuroscience* 8 (2014). www.ncbi.nlm.nih.gov/pmc/articles/PMC3902298.

CHAPTER 13. CO-CREATORS

1. R. Rosenthal and L. Jacobson, "Teachers' Expectancies: Determinants of Pupils' IQ Gains," *Psychological Reports* 19 (1963): 115–18.
2. Graham Lawton, "The Grand Delusion: Why Nothing Is as It Seems," *New Scientist* May 6, 2011. www.learningmethods.com/downloads/pdf/the.grand.delusion-why.nothing.is.as.it.seems.pdf.
3. Paul Piff, "Does Money Make You Mean?" TED Talk, October 2013. www.ted.com/talks/paul_piff_does_money_make_you_mean?language=en.
4. B. G. Dias and K. J. Ressler, "Parental Olfactory Experience Influences Behavior and Neural Structure in Subsequent Generations," *Nature Neuroscience* 17 (2014): 89–96. http://dx.doi.org/10.1038/nn.3594.
5. R. Yehuda, "Neuroendocrine Aspects of PTSD," *Handbook of Experimental Pharmacology* 169 (2005): 371–403. www.ncbi.nlm.nih.gov/pubmed/16594265.
6. R. Yehuda, N. P. Daskalais, H. N. Bierer, T. Klengel, F. Holsboer, and E. B. Binder, "Holocaust Exposure Induced Intergenerational

Effects on FKBP5 Methylation," *Biological Psychiatry* (2015). www.biologicalpsychiatryjournal.com/article/S0006-3223(15)00652-6/abstract.

7. R. Boyd and P. J. Richerson, *Not by Genes Alone: How Culture Transformed Human Evolution* (Chicago: University of Chicago Press, 2006).
8. S. Bowles and H. Gintis, "The Evolution of Strong Reciprocity: Cooperation in Heterogeneous Populations," *Theoretical Population Biology* 65, no. 1 (February 2004): 17–28. www.umass.edu/preferen/gintis/evolsr.pdf.
9. N. A. Christakis and J. H. Fowler, *Connected: The Surprising Power of Our Social Networks and How They Shape Our Lives—How Your Friends' Friends' Friends Affect Everything You Feel, Think, and Do* (New York: Back Bay Books/Little Brown, 2011).
10. K. Picknett and R. Wilkinson, *The Spirit Level: Why Greater Equality Makes Societies Stronger* (New York: Bloomsbury, 2011).

CHAPTER 14. CONSCIOUS EVOLUTION

1. S. Grof, *When the Impossible Happens: Adventures in Non-Ordinary Realities* (Louisville, Colo.: Sounds True, 2006).

Index